The Number System

Coins, Coupons, and Combinations

Grade 2

Also appropriate for Grade 3

Karen Economopoulos
Susan Jo Russell

Developed at TERC, Cambridge, Massachusetts

Dale Seymour Publications®

The *Investigations* curriculum was developed at TERC (formerly
Technical Education Research Centers) in collaboration with Kent State
University and the State University of New York at Buffalo. The work was
supported in part by National Science Foundation Grant No. ESI-9050210.
TERC is a nonprofit company working to improve mathematics and
science education. TERC is located at 2067 Massachusetts Avenue,
Cambridge, MA 02140.

**This project was supported, in part,
by the
National Science Foundation**
Opinions expressed are those of the authors
and not necessarily those of the Foundation

This book is published by Dale Seymour Publications®, an imprint of the
Alternative Publishing Group of Addison-Wesley Publishing Company.

Editorial Management and Production: McClanahan and Company
Managing Editor: Catherine Anderson
Project Editor: Alison Abrohms
Series Editor: Beverly Cory
ESL Consultant: Nancy Sokol Green
Production/Manufacturing Director: Janet Yearian
Production/Manufacturing Coordinator: Shannon Miller
Design Manager: Jeff Kelly
Design: Don Taka
Composition: Book Publishing Enterprises, Inc.
Illustrations: Meryl Treatner, Laurie Harden, Barbara Epstein-Eagle
Cover: Bay Graphics

Printed on Recycled Paper

**DALE
SEYMOUR
PUBLICATIONS®**
P.O. BOX 10888
PALO ALTO, CA 94303

Order number DS21646
ISBN 1-57232-215-2
1 2 3 4 5 6 7 8 9 10-ML-00 99 98 97 96

TERC

INVESTIGATIONS IN NUMBER, DATA, AND SPACE

Principal Investigator   Susan Jo Russell

Co-Principal Investigator   Cornelia C. Tierney

Director of Research and Evaluation   Jan Mokros

Director of K–2 Curriculum   Karen Economopoulos

Curriculum Development
Joan Akers
Michael T. Battista
Mary Berle-Carman
Douglas H. Clements
Karen Economopoulos
Anne Goodrow
Marlene Kliman
Jerrie Moffett
Megan Murray
Ricardo Nemirovsky
Andee Rubin
Susan Jo Russell
Cornelia C. Tierney
Tracey Wright

Evaluation and Assessment
Mary Berle-Carman
Jan Mokros
Andee Rubin

Teacher Support
Anne Goodrow
Liana Laughlin
Jerrie Moffett
Megan Murray
Tracey Wright

Technology Development
Michael T. Battista
Douglas H. Clements
Julie Sarama

Video Production
David A. Smith
Judy Storeygard

Administration and Production
Irene Baker
Amy Catlin
Amy Taber

Cooperating Classrooms for This Unit
Rose Christiansen
Brookline Public Schools, Brookline, MA
Lisa Seyferth
Carol Walker
Newton Public Schools, Newton, MA
Phyllis Ollove
Boston Public Schools, Boston, MA
Margaret M. McGaffigan
Nashoba Regional School District, Stow, MA
Marina Seevac
Cambridge Public School, Cambridge, MA
Barbara Rynerson
Dale Dhoore
Beth Newkirk
Pat McLure
Oyster River Public Schools, Durham, NH

Consultants and Advisors
Deborah Lowenberg Ball
Marilyn Burns
Ann Grady
James J. Kaput
Mary M. Lindquist
John Olive
Leslie P. Steffe
Grayson Wheatley

Graduate Assistants
Kent State University:
Kathryn Battista, Caroline Borrow, Judy Norris

State University of New York at Buffalo:
Julie Sarama, Sudha Swaminathan,
Elaine Vukelic

CONTENTS

Teacher Notes

Investigations in Number, Data, and Space is a K–5 mathematics curriculum with four major goals:

- to offer students meaningful mathematical problems
- to emphasize depth in mathematical thinking rather than superficial exposure to a series of fragmented topics
- to communicate mathematics content and pedagogy to teachers
- to substantially expand the pool of mathematically literate students

The *Investigations* curriculum embodies an approach radically different from the traditional textbook-based curriculum. At each grade level, it consists of a set of separate units, each offering 2–6 weeks of work. These units of study are presented through investigations that involve students in the exploration of major mathematical ideas.

Approaching the mathematics content through investigations helps students develop flexibility and confidence in approaching problems, fluency in using mathematical skills and tools to solve problems, and proficiency in evaluating their solutions. Students also build a repertoire of ways to communicate about their mathematical thinking, while their enjoyment and appreciation of mathematics grow.

The investigations are carefully designed to invite all students into mathematics—girls and boys; diverse cultural, ethnic, and language groups; and students with different strengths and interests. Problem contexts often call on students to share experiences from their family, culture, or community. The curriculum eliminates barriers—such as work in isolation from peers, or emphasis on speed and memorization—that exclude some students from participating successfully in mathematics. The following aspects of the curriculum ensure that all students are included in significant mathematics learning.

- Students spend time exploring problems in depth.
- They find more than one solution to many of the problems they work on.
- They invent their own strategies and approaches, rather than relying on memorized procedures.
- They choose from a variety of concrete materials and appropriate technology, including calculators, as a natural part of their everyday mathematical work.
- They express their mathematical thinking through drawing, writing, and talking.
- They work in a variety of groupings—as a whole class, individually, in pairs, and in small groups.
- They move around the classroom as they explore the mathematics in their environment and talk with their peers.

While reading and other language activities are typically given a great deal of time and emphasis in elementary classrooms, mathematics often does not get the time it needs. If students are to experience mathematics in depth, they must have enough time to become engaged in real mathematical problems. We believe that a minimum of 5 hours of mathematics classroom time a week—about an hour a day—is critical at the elementary level. The plan and pacing of the *Investigations* curriculum are based on that belief.

For further information about the pedagogy and principles that underlie these investigations, see the Teacher Notes throughout the units and the following books:

- *Implementing the* Investigations in Number, Data, and Space® *Curriculum*
- *Beyond Arithmetic: Changing Mathematics in the Elementary Classroom*

The *Investigations* curriculum is presented through a series of teacher books, one for each unit of study. These books not only provide a complete mathematics curriculum for your students, they also offer materials to support your own professional development. You, the teacher, are the person who will make this curriculum come alive in the classroom; the book for each unit is your main support system.

While reproducible resources for students are provided, the curriculum does not include student books. Students work actively with objects and experiences in their own environment and with a variety of manipulative materials and technology, rather than with workbooks and worksheets filled with problems. We also make extensive use of the overhead projector as a way to present problems, to focus group discussion and to help students share ideas and strategies. If an overhead projector is available, we urge you to try it as suggested in the investigations.

Ultimately, every teacher will use these investigations in ways that make sense for his or her particular style, the particular group of students, and the constraints and supports of a particular school environment. We have tried to provide with each unit the best information and guidance for a wide variety of situations, drawn from our collaborations with many teachers and students over many years. Our goal in this book is to help you, as a professional educator, implement this mathematics curriculum in a way that will give all your students access to mathematical power.

Investigation Format

The opening two pages of each investigation help you get ready for the student work that follows. Here you will read:

What Happens—a synopsis of each session or block of sessions.

Mathematical Emphasis—the most important ideas and processes students will encounter in this investigation.

What to Plan Ahead of Time—materials to gather, student sheets to duplicate, transparencies to make, and anything else you need to do before starting.

INVESTIGATION 3

Introducing Addition and Subtraction Situations

What Happens

Session 1: Introducing Combining Situations
Students are introduced to combining problems. They solve problems and record their solution so that someone else can understand it. Several strategies are shared and recorded.

Session 2: Combining Notation In pairs, students share the combining problem they wrote for homework. As a class they make up problem situations for the expression 21 + 14. For most of the session they work on solving combining situations and writing a problem represented by combining notation.

Session 3: Introducing Separating Situations
Students are introduced to separating problems. Similar to the previous sessions, they discuss a story problem, solve a few separating problems, and record their solution so that someone else can understand it. Several strategies are shared and recorded.

Sessions 4 and 5: Making Sense of Addition and Subtraction Students solve a variety of story problems involving combining and separating, using their own strategies. Their job is to solve each problem, check their solutions, and clearly record their approaches. As a whole class, students discuss the relationship between addition and subtraction situations.

Mathematical Emphasis

■ Developing models of addition and subtraction situations
■ Solving problems using numerical reasoning
■ Recording solution strategies clearly
■ Considering the relationship between addition and subtraction
■ Working with notation for addition and subtraction
■ Matching addition and subtraction notation to situations they could represent

INVESTIGATION 3

What to Plan Ahead of Time

Materials

■ Overhead projector (Sessions 1 and 3, optional)
■ Counters, such as interlocking cubes or tiles (Sessions 1–5)
■ Envelopes: 7 (Sessions 4 and 5)
■ Paste or glue sticks (Sessions 4 and 5)

Other Preparation

For Session 1
■ Think about situations familiar to students that you might use as contexts for addition and subtraction problems. Throughout this unit, you may want to substitute problems of your own. For more information, see the **Teacher Note**, Creating Your Own Story Problems (p. 108). If you use your own problem contexts, create the following kinds of problems: a combining situation (with one or two follow-up problems) for Sessions 1 and 2, a separating situation (with one or two follow-up problems) for Session 3, and a mixture of six to eight combining and separating problems for Sessions 4 and 5.
■ Duplicate Student Sheet 18, Story Problems, Set A (or copies of the problems you have created), 1 per student.

For Session 2
■ Duplicate Student Sheet 19, Story Problems, Set B (or copies of the problems you have created), 1 per student.

For Session 3
■ Duplicate Student Sheet 20, Story Problems, Set C (or copies of the problems you have created), 1 per student.

For Sessions 4 and 5
■ Set D (or copies of the problems you have created), 1 per student.
■ Prepare story problems on Student Sheet 21 by cutting apart the sheet into individual problems. Store each problem in a separate envelope marked with its number. Paste an example of each problem on the front of the envelope so students can see which problem they are choosing.

Sessions Within an investigation, the activities are organized by class session, a session being a 1-hour math class. Sessions are numbered consecutively throughout an investigation. Often several sessions are grouped together, presenting a block of activities with a single major focus.

When you find a block of sessions presented together—for example, Sessions 1, 2, and 3—read through the entire block first to understand the overall flow and sequence of the activities. Make some preliminary decisions about how you will divide the activities into three sessions for your class, based on what you know about your students. You may need to modify your initial plans as you progress through the activities, and you may want to make notes in the margins of the pages as reminders for the next time you use the unit.

Be sure to read the Session Follow-Up section at the end of the session block to see what homework assignments and extensions are suggested as you make your initial plans.

While you may be used to a curriculum that tells you exactly what each class session should cover, we have found that the teacher is in a better position to make these decisions. Each unit is flexible and may be handled somewhat differently by every teacher. While we provide guidance for how many sessions a particular group of activities is likely to need, we want you to be active in determining an appropriate pace and the best transition points for your class.

Start-Up The Start-Up section at the beginning of each session offers suggestions for how to acknowledge and integrate homework from the previous session and which Classroom Routine activities to include sometime during the school day.

Classroom Routines Routines provide students regular practice in important mathematical skills such as solving number combinations, collecting and organizing data, and understanding time. There are three classroom routines that occur regularly throughout the grade 2 *Investigations* curriculum, and a fourth one that occurs in the Geometry and Fractions unit, *Shapes, Halves, and Symmetry*.

Session 1

Introducing Combining Situations

Materials

- Counters
- Student Sheet 18 (1 per student)
- Overhead projector (optional)

What Happens

Students are introduced to combining problems. They solve a problem and record their solution so that someone else can understand it. Several strategies are shared and recorded. Their work focuses on:

- combining quantities
- recording strategies

Start-Up

Today's Number Suggest that students use combinations of 10 in their number sentences. For example, if the number they are working on is 54 and one number sentence is 10 + 10 + 10 + 10 + 4, ask students if there is another way of making 10, such as 6 + 4 + 6 + 4 + 6 + 4 + 6 + 4 + 6 + 4 + 4. Add a card to the class counting strip and fill in another number on the blank 200 chart.

Activity

Problems About Combining

As students are introduced to story problems, try to refrain from labeling them ahead of time as "addition" or "subtraction." See the **Teacher Note**, The Relationship Between Addition and Subtraction (p. 118), for more information on ways that students interpret problems. For these five sessions, students will be presented with addition problems that involve finding a total amount by combining two quantities and subtraction problems that involve starting with a whole quantity and then removing or separating a part of that quantity. These ideas are expanded upon in a later unit, *Putting Together and Taking Apart*, in which students will be introduced to a variety of subtraction situations including comparing two quantities.

Write a problem such as the one that follows on the chalkboard or overhead. Keeping the same numbers and the basic structure of the problem, you might want to create a problem that has a more familiar or timely context for your students. See the **Teacher Note**, Creating Your Own Story Problems (p. 108).

104 ▪ *Investigation 3: Introducing Addition and Subtraction Situations*

One of two classroom routines, Today's Number or How Many Pockets?, is integrated into the Start-Up of most sessions. The third routine, Time and Time Again appears as a Start-Up in the final unit, *Timelines and Rhythm Patterns*, but is not integrated directly into other units. Instead it is offered as a resource of activities about understanding time and the passage of time. This routine can be integrated throughout the school day and into other parts of the classroom curriculum.

Classroom routines offer students opportunities to build on a familiar activity by integrating experiences from previously taught units. For example, in the routine Today's Number, students write number sentences that equal the number of days they have been in school. Variations of this routine include using addition and subtraction to express the number, using multiples of 5 and 10 to express the number, and using more than three addends.

Most classroom routine activities are short and can be done whenever you have a spare 10 minutes— maybe before lunch or recess, or at the beginning or end of the day. Complete descriptions of each Classroom Routine can be found at the end of each unit.

Activities   The activities include pair and small-group work, individual tasks, and whole-class discussions. Students are seated together, talking and sharing ideas during all work times. Students most often work cooperatively, although each student may record work individually.

Choice Time   In most units, there are sessions that are structured with activity choices. In these cases, students may work simultaneously on different activities focused on the same mathematical ideas. Students choose which activities they want to do and they cycle through them.

You will need to decide how to set up and introduce these activities and how to let students make their choices. Some teachers present them as station activities in different parts of the room. Other teachers post a list of activities and have students collect their own materials and choose their own work space. You may need to experiment with a few different structures before finding one that works best for you and your students.

Tips for the Linguistically Diverse Classroom

At strategic points in each unit, you will find concrete suggestions for simple modifications of the teaching strategies to encourage the participation of all students. Many of these tips offer alternative ways to elicit critical thinking from students at varying levels of English proficiency, as well as from other students who find it difficult to verbalize their thinking.

The tips are supported by suggestions for specific vocabulary work to help ensure that all students can participate fully in the investigations. The Preview for the Linguistically Diverse Classroom (p. 15) lists important words that are assumed as part of the working vocabulary of the unit. Second-language learners will need to become familiar with these words in order to understand the problems and activities they will be doing. These terms can be incorporated into students' second-language work before or during the unit. Activities that can be used to present the words are found in the appendix, Vocabulary Support for Second-Language Learners (p. 150).

In addition, ideas for making connections to students' language and cultures, included on the Preview page, help the class explore the unit's concepts from a multicultural perspective.

Session Follow-Up

Homework   Homework is not given daily for its own sake, but periodically as it makes sense to have follow-up work at home. Homework may be used for (1) review and practice of work done in class; (2) preparation for activities coming up—for example, collecting data for a class project; or (3) involving and informing family members.

Some units in the *Investigations* curriculum have more homework than others, simply because it makes sense for the mathematics that's going on. Other units rely on manipulatives that most students won't have at home, making homework difficult. In any case, homework should always be directly connected to the investigations in the unit or to work in previous units—never sheets of problems just to keep students busy.

Extensions   These follow-up activities are opportunities for some or all students to explore a topic in greater depth or in a different context. They are not designed for "fast" students; mathematics is a multi-faceted discipline, and different students will want to go further in different investigations. Look for and encourage the sparks of interest and enthusiasm you see in your students and use the extensions to help them pursue these interests.

Family Letter   A letter that you can send home to students' families is included with the blackline masters for each unit. We want families to be informed about the mathematics work in your classroom; they should be encouraged to participate in and support their children's work. A reminder to send home the letter appears in one of the early investigations. These letters are also available separately in Spanish, Vietnamese, Cantonese, Hmong, and Cambodian.

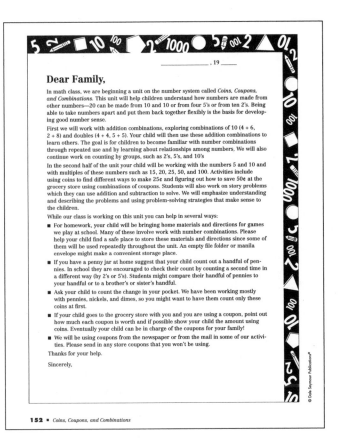

each unit for how to organize different types of computer environments.

Children's Literature

Every unit in the grade 2 *Investigations* curriculum offers a suggested bibliography of children's literature that can be used to support the mathematical ideas presented in the unit. This bibliography is found on the Materials List located in the front of each unit.

Some of the grade 2 units have class sessions that are based on a selected children's book. Although the session can be taught without a book, using it offers a rich introduction to the activity.

Literature selected to include in both the units and the bibliographies was limited to books that could offer students strong connections to the mathematics they will be investigating.

Technology

Calculators and computers are important components of the *Investigations* curriculum.

Calculators are introduced to students in the second unit, *Coins, Coupons, and Combinations*, of the grade 2 sequence. It is assumed that calculators are a readily available material throughout the curriculum.

The grade 2 *Investigations* curriculum uses two software programs that were developed especially for the curriculum. *Shapes* is introduced in the Introductory unit, *Mathematical Thinking at Grade 2,* and used again during the Geometry and Fractions unit, *Shapes, Halves, and Symmetry.* *Geo-Logo* is introduced and used in the Measurement unit, *How Long? How Far?* Although the software is included in only these three units, we recommend that students use it throughout the year. As students use the software over time, they continue to develop skills presented in the units.

How you incorporate the computer activities into your curriculum depends on the number of computers you have available. Suggestions are offered in

Materials

A complete list of the materials needed for the unit is found on p. 13. Some of these materials are available in a kit for the grade 2 *Investigations* curriculum. Individual items can also be purchased as needed from school supply stores and dealers.

In an active mathematics classroom, certain basic materials should be available at all times: interlocking cubes, pencils, unlined paper, graph paper, calculators, and things to count with. Some activities in this curriculum require scissors and glue sticks or tape. Stick-on notes and large chart paper are also useful materials throughout.

So that students can independently get what they need at any time, they should know where these materials are kept, how they are stored, and how they are to be returned to the storage area. Many teachers have found that stopping 5 minutes before the end of each session so that students can finish what they are working on and clean up is helpful in maintaining classroom materials. You'll find that establishing such routines at the beginning of the year is well worth the time and effort.

Student Sheets

Reproducible pages to help you teach the unit are found at the end of this book. These include student recording sheets as well as masters that can be used as teaching tools.

Many of the field-test teachers requested more sheets to help students record their work, and we have tried to be responsive to this need. At the same time, we think it's important that students find their own ways of organizing and recording their work. They need to learn how to explain their thinking with both drawings and written words, and how to organize their results so someone else can understand them.

To ensure that students get a chance to learn how to represent and organize their own work, we deliberately do not provide student sheets for every activity. We recommend that students keep a folder so that their work, whether on reproducible sheets or their own paper, is always available to them for reference.

Name _____ Date _____

Student Sheet 10

Close to 20 Score Sheet

GAME 1 SCORE

Round 1: _____ + _____ + _____ = _____   _____

Round 2: _____ + _____ + _____ = _____   _____

Round 3: _____ + _____ + _____ = _____   _____

Round 4: _____ + _____ + _____ = _____   _____

Round 5: _____ + _____ + _____ = _____   _____

TOTAL SCORE   _____

GAME 2 SCORE

Round 1: _____ + _____ + _____ = _____   _____

Round 2: _____ + _____ + _____ = _____   _____

Round 3: _____ + _____ + _____ = _____   _____

Round 4: _____ + _____ + _____ = _____   _____

Round 5: _____ + _____ + _____ = _____   _____

TOTAL SCORE   _____

162 ■ *Coins, Coupons, and Combinations*

Help for You, the Teacher

Because we believe strongly that a new curriculum must help teachers think in new ways about mathematics and about their students' mathematical thinking processes, we have included a great deal of material to help you learn more about both.

About the Mathematics in This Unit   This introductory section (p. 14) summarizes the critical information about the mathematics you will be teaching. This will be particularly valuable to teachers who are accustomed to a traditional textbook-based curriculum.

Teacher Notes   These reference notes provide practical information about the mathematics you are teaching and about our experience with how students learn. Many of the notes were written in response to actual questions from teachers, or to discuss important things we saw happening in the field-test classrooms. Some teachers like to read them all before starting the unit, then review them as they come up in particular investigations.

Dialogue Boxes   Sample dialogues throughout the unit demonstrate how students typically express their mathematical ideas, what issues and confusions arise in their thinking, and how some teachers have guided class discussions.

These dialogues are based on the extensive classroom testing of this curriculum; many are word-for-word transcriptions of recorded class discussions. The value of these dialogues is that they offer good clues to how students may develop and express their approaches and strategies, helping you prepare for your own class discussions.

Where to Start   You may not have time to read everything the first time you use this unit. As a first-time user, you will likely focus on understanding the activities and working them out with your students. Read completely through each investigation before starting to present it.

When you next teach this same unit, you can begin to read more of the background. Each time you present this unit, you will learn more about how students understand the mathematical ideas. The first-time user of *Coins, Coupons, and Combinations* should read the following:

- About the Mathematics in This Unit (p. 14)
- Teacher Note: Strategies for Learning Addition Combinations (p. 46)
- Teacher Note: Students' Addition and Subtraction Strategies (p. 109)
- Teacher Note: Developing Numerical Strategies (p. 110)
- Teacher Note: The Relationship Between Addition and Subtraction (p. 118)

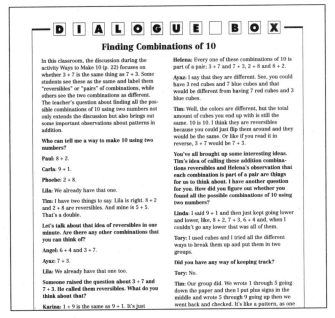

Observing the Students

Throughout the *Investigations* curriculum there are numerous opportunities to observe students as they work. Teacher observations are an important part of ongoing assessment. Individual observations provide snapshots of a student's experience with a single activity. A series of observations over time can provide an informative and detailed picture of a student that is useful in documenting and assessing his or her growth. They are important sources of information when preparing for family conferences or writing student reports.

For many activities, we offer observation guidelines and suggestions. These include what to look for as students work, or questions you can ask to give you insights into their thinking or to stimulate further exploration of an activity.

Teacher Checkpoints The Teacher Checkpoints provide a time for you to pause and reflect on your teaching plan, observe students at work, and get an overall sense of how your class is doing in the unit. These sections offer tips on what you should be looking for and how you might adjust your pacing. Are most students fluent with strategies for solving a particular kind of problem? Are they just starting to formulate good strategies? Or are they still struggling with how to start?

Depending on what you see as students work, you may want to spend more time on similar problems, change some of the problems to use smaller numbers, move quickly to more challenging material, modify subsequent activities for some students, work on particular ideas with a small group, or pair students who have good strategies with those who are having more difficulty.

In *Coins, Coupons, and Combinations* you will find these Teacher Checkpoints:

> How Many Fingers? (p. 87)
> Student Strategies (p. 121)

Embedded Assessment Activities Use the built-in assessments included in this unit to help you examine the work of individual students, figure out what it means, and provide feedback. From the students' point of view, the activities you will be using for assessment are no different from any others; they don't look or feel like traditional tests.

These activities sometimes involve writing and reflecting, a brief interaction between student and teacher, or the creation and explanation of a product.

In *Coins, Coupons, and Combinations* you will find these assessment activities:

> Our Class and the Magic Pot (p. 63)
> Collecting Pocket Data (p. 137)

Teachers find that the hardest part of the assessment is interpreting students' work. If you have used a process approach to teaching writing, you will find our mathematics approach familiar. To help with interpretation, we provide guidelines and questions to ask about students' work. In many cases a Teacher Note with specific examples of student work and a commentary on what they indicate is included. This framework can help you determine how students are progressing.

As you evaluate students' work, it's important to remember that you're looking for much more than the "right answer." You'll want to know what their strategies are for solving the problem, how well these strategies work, whether they can keep track of and logically organize an approach to the problem, and how they make use of representations and tools to solve the problem.

Ongoing Assessment   Good assessment of student work involves a combination of approaches. Some of the things you might do on an ongoing basis include the following:

- **Observation**   Circulate around the room to observe students as they work. Watch for the development of their mathematical strategies and listen to their discussions of mathematical ideas.

- **Portfolios**   Ask students to document their work in journals, notebooks, or portfolios. Periodically review this work to see how their mathematical thinking and writing are changing. Some teachers have students keep a notebook or folder for each unit, while others prefer one mathematics notebook or a portfolio of selected work for the entire year. Take time at the end of each unit for students to choose work for their portfolios. You might also have them write about what they've learned in the unit.

Coins, Coupons, and Combinations

OVERVIEW

Content of This Unit The focus of this unit is on helping students develop a sense of numbers as whole quantities and to begin to look for patterns and relationships that exist in our number system. Students gain facility with addition combinations through their work with combinations of 10, doubles (4 + 4), and doubles plus or minus 1 (4 + 5, 4 + 3). They continue their work with counting, with an emphasis on counting by groups, such as by 2's, 5's, and 10's, and use these counting strategies to keep track of large amounts. Students explore 5 and 10 and multiples of 5 and 10 and begin to look for relationships between these important numbers in our number system. They are introduced to combining and separating problems and are encouraged to develop strategies that make sense to them for adding and subtracting numbers. In addition, students are introduced to calculators, coins, and the 100 chart and are encouraged to use these tools as they solve problems.

Connections with Other Units If you are doing the full-year *Investigations* curriculum in the suggested sequence for grade 2, this is the second of eight units. The work in this unit is an extension of the counting and number work introduced in *Mathematical Thinking at Grade 2*. The concepts and ideas that are developed in this unit are extended in the addition and subtraction unit, *Putting Together and Taking Apart*.

This unit can also be used successfully at grade 3, depending on the previous experience and needs of your students.

Investigations Curriculum ■ Suggested Grade 2 Sequence

Mathematical Thinking at Grade 2 (Introduction)

▶ *Coins, Coupons, and Combinations* (The Number System)

Does It Walk, Crawl, or Swim? (Sorting and Classifying Data)

Shapes, Halves, and Symmetry (Geometry and Fractions)

Putting Together and Taking Apart (Addition and Subtraction)

How Long? How Far? (Measurement)

How Many Pockets? How Many Teeth? (Collecting and Representing Data)

Timelines and Rhythm Patterns (Representing Time)

Investigation 1 • 10's and Doubles		
Class Sessions	**Activities**	**Pacing**
Session 1 (page 21) COMBINATIONS OF 10	Ways to Make 10 *Book of 10* Introducing Math Folders and Weekly Logs	1 hr
Sessions 2 and 3 (page 29) CARD GAMES	Tens Go Fish and Turn Over 10 Introducing Choice Time ■ Homework	2 hrs
Sessions 4 and 5 (page 38) DOUBLES	Two of Everything Book of Magic Pot Riddles Sharing Magic Pot Riddles ■ Homework	2 hrs
Session 6 (page 43) NUMBER STRINGS	Strategies for Combining Two Numbers ■ Homework	1 hr
Session 7 (page 48) EXPLORING CALCULATORS	Exploring Calculators Problems to Solve on the Calculator Sharing Problems	1 hr
Sessions 8 and 9 (page 55) CLOSE TO 20 AND BEAT THE CALCULATOR	Close to 20 Beat the Calculator Choice Time Class Discussion: Combinations Related to Doubles ■ Homework	2 hrs
Session 10 (page 61) TODAY'S NUMBER AND THE MAGIC POT	Today's Number: Using Number Combinations We Know ■ Assessment: Our Class and the Magic Pot	1 hr
Session 11 (page 66) COUNTING POCKETS	Counting Pockets Finishing Choices	1 hr

Investigation 2 • Grouping by 2's, 5's, and 10's		
Class Sessions	**Activities**	**Pacing**
Session 1 (page 72) USING GROUPS OF 2	How Many Legs Are in Our Class? Looking for Patterns ■ Homework	1 hr
Session 2 (page 78) EXPLORING MULTIPLES OF 5	Solving People and Pet Riddles Counting Around the Class	1 hr
Session 3 (page 81) COUNTING BY DIFFERENT GROUPS	Today's Number Counting Bags	1 hr

Continued on next page

Class Sessions	Activities	Pacing
Sessions 4 and 5 (page 85) COUNTING CHOICES	Choice Time ■ Teacher Checkpoint: How Many Fingers?	2 hrs
Session 6 (page 92) WAYS TO MAKE 15¢	Ways to Make 15¢ Matching Coins and Coupons ■ Homework	1 hr
Sessions 7, 8, and 9 (page 95) COINS AND COUPONS	Collect 50¢ and Shop and Save Choice Time Class Discussion: Ways to Make 25¢	3 hrs
Session 10 (page 99) COLLECTING POCKET DATA	Counting Pockets Finishing Choices	1 hr

Investigation 3 • Introducing Addition and Subtraction Situations

Class Sessions	Activities	Pacing
Session 1 (page 104) INTRODUCING COMBINING SITUATIONS	Problems About Combining Sharing Strategies ■ Homework	1 hr
Session 2 (page 112) COMBINING NOTATION	Looking at Addition Notation	1 hr
Session 3 (page 115) INTRODUCING SEPARATING SITUATIONS	Problems About Separating Sharing Strategies ■ Homework	1 hr
Sessions 4 and 5 (page 119) MAKING SENSE OF ADDITION AND SUBTRACTION	Looking at Subtraction Notation Story Problems ■ Teacher Checkpoint: Student Strategies Class Discussion: Is It Adding or Subtracting?	2 hrs

Investigation 4 • One Hundred

Class Sessions	Activities	Pacing
Session 1 (page 126) EXPLORING THE 100 CHART	Filling in the 100 Chart What Do You Notice About the 100 Chart?	1 hr
Sessions 2, 3, and 4 (page 129) WORKING WITH 100	Today's Number: How Far from 100? Roll-a-Square Choice Time Class Discussion: Many Ways to Solve a Problem ■ Homework	3 hrs
Session 5 (page 137) PENNY-A-POCKET	■ Assessment: Collecting Pocket Data Choosing Student Work to Save	1 hr

Following are the basic materials needed for the activities in this unit. The suggested quantities are ideal; however, in some instances you can work with smaller quantities by running several activities, requiring different materials, simultaneously.

Items marked with an asterisk are provided with the *Investigations* Materials Kit for grade 2.

* Interlocking cubes: about 30 per student

* Hundred Number Wall Chart with number cards 1–100 and transparent pattern markers

* Plastic coin sets (real coins may be substituted), 30 pennies, 20 nickels, 20 dimes: 1 set per 3–4 students

  Containers for coin sets

* Primary Number Cards (optional), referred to as Number Cards throughout the unit (manufactured); or use blackline masters to make your own cards: 1 set per student

  Calculators: at least 1 per pair, ideally 1 per student

* Number cubes (with numbers or dots 1–6): 2 per 2–3 students

  Two of Everything by Lily Toy Hong. Morton Grove, Ill.: A. Witman, 1993. (optional)

  Chart paper or newsprint

  5"-by-8" cards or sheets of paper: at least 1 per student

  Clipboards: 1 per student (optional)

  Construction paper, white or light color, 12"-by-18"

  Crayons or markers

  Folders for student work: 1 per student

  Empty shoe box or container

  Resealable plastic bags or envelopes for storing paper cards: 2 per student

  Large envelopes for storing story problems: about 14

  Paste or glue sticks

  Large jar

  Overhead projector (optional)

  Collection of coupons from newspaper or store flyers

  Magic Pot items such as a dime, a box of markers, a bag containing 6–8 cubes

  10–15 bags of assorted small objects, such as pennies, beans, cubes (each bag should be lettered for identification purposes): 1 per pair plus a few extra

* Counters (buttons, pennies, beans)

The following materials are provided at the end of this unit as blackline masters.

Family letter (p. 152)
Student Sheets 1–26 (pp. 153–178)

Suggested Bibliography of Children's Literature

Bogart, Jo Ellen. *10 for Dinner*. New York: Scholastic Inc., 1989.

Friedman, Aileen. *The King's Commissioners*. New York: Scholastic Inc., 1994.

Hamm, Diane Johnson. *How Many Feet in the Bed?* New York: Simon and Schuster, 1991.

Hong, Lily Toy. *Two of Everything*. Morton Grove, Ill A. Witman, 1993.

Pinczes, Elinor J. *One Hundred Hungry Ants*. Boston: Houghton Mifflin, 1993.

An important part of the mathematical work that students do in the primary grades is focused on developing a strong understanding of numbers and number composition. This understanding provides students with a firm foundation for understanding the base-ten number system. Good number sense is central to much of the work students will do in mathematics. Throughout this unit students are provided with opportunities and experiences to help them learn about the structure of numbers, number relationships, and the patterns that exist in our number system.

Counting is central to most of the number work that primary-age students do. Counting by 1's involves not only knowing the rote counting sequence but also being able to keep track of what has been counted. Eventually students are able to move toward counting by groups, understanding that one number can represent a group of many objects. It is from here that students begin to build an understanding of numbers and how they are composed. Being able to take numbers apart into useful parts, manipulate these parts, and then put them back together reflect the sort of flexible thinking that leads to powerful mathematical knowledge.

Flexible thinking in the form of pulling numbers apart and putting them back together is often reflected in students' ability to develop effective strategies for adding and subtracting numbers. Before students are able to do this, they must understand the problem and be able to visualize or build a representation of what is happening in the problem. Thus the focus is not only on being able to add and subtract numbers but also on being able to understand what is involved in both of those operations.

Mathematical Emphasis   At the beginning of each investigation, the Mathematical Emphasis section tells you what is most important for students to learn about during that investigation. Many of these mathematical understandings and processes are difficult and complex. Students gradually learn more and more about each idea over many years of schooling. Individual students will begin and end the unit with different levels of knowledge and skill, but all will gain greater knowledge about counting and structure of numbers and will use this knowledge as a basis for understanding addition and subtraction situations.

In the *Investigations* curriculum, mathematical vocabulary is introduced naturally during the activities. We don't ask students to learn definitions of new terms; rather, they come to understand such words as *factor* or *area* or *symmetry* by hearing them used frequently in discussion as they investigate new concepts. This approach is compatible with current theories of second-language acquisition, which emphasize the use of new vocabulary in meaningful contexts while students are actively involved with objects, pictures, and physical movement.

Listed below are some key words used in this unit that will not be new to most English speakers at this age level but may be unfamiliar to students with limited English proficiency. You will want to spend additional time working on these words with students who are learning English. If students are working with a second-language teacher, you might enlist your colleague's aid in familiarizing students with these words before and during this unit. In the classroom, look for opportunities for students to hear and use these words. Activities you can use to present the words are given in the appendix, Vocabulary Support for Second-Language Learners (p. 150).

imagine, story problem   Students need recognize the action in a story problem to be able to tell whether the situation involves a joining or separating of groups or sets. To help students with this, they are often asked to imagine a scene or scenerio.

pennies, nickels, dimes, coins, group   Students learn to identify U.S. coins and recognize the value of each as they work with problems involving money, including combining and comparing different amounts. A group (or set) of coins will be the phrase that is used to refer to several coins in a set.

Multicultural Extensions for All Students

Whenever possible, encourage students to share words, objects, customs, or any aspects of daily life from their own cultures and backgrounds that are relevant to the activities in this unit. For example,

- As students find combinations for 10 in Investigation 1, they may be interested in finding out how the number 10 is expressed in various cultures and languages. Invite students to share what they know and to do research on how the number 10 is written and spoken in various cultures and languages.

- As students use a calculator to solve number string problems in Investigation 1, they may enjoy finding out about mathematical tools in various cultures. Invite students to do research on these tools, such as the Chinese abacus, and to find out how to use the tools for mathematics.

- Students will play card games involving the number 10 in Investigation 1. They may know of games involving numbers from their own cultures or other cultures. Invite students to share the games they know or to do research on games of various cultures.

Investigations

10's and Doubles

What Happens

Session 1: Combinations of 10 Students work with combinations of 10. They list ways of making 10 with two and three numbers. They make a *Book of 10*, listing all of the possible combinations of 10 using two to ten addends.

Sessions 2 and 3: Card Games Students are introduced to two card games, Tens Go Fish and Turn Over 10. These, along with *Book of 10*, are choices for the next two sessions. (Note: If students are familiar with these games, these sessions are optional.)

Sessions 4 and 5: Doubles The book *Two of Everything* introduces students to the concept of doubling. They generate a list of items and then figure out how many they would have if they were doubled. As a class they make a book of doubles riddles.

Session 6: Number Strings Students discuss strategies for adding "strings" of numbers. They use doubles (3 + 3, 4 + 4) and combinations of 10 (3 + 7, 4 + 6) to solve other addition problems. They record their solutions and then discuss their strategies with partners.

Session 7: Exploring Calculators Students explore the calculator as a tool for mathematics. As they make up and solve problems on the calculator, you have a chance to see how familiar they are with this tool. In a class discussion, students share what they know about the calculator, including use of the operation keys and an awareness of the decimal point. As a class they solve number string problems.

Sessions 8 and 9: Close to 20 and Beat the Calculator Students are introduced to two activities that involve adding two or more numbers. For two sessions they work on both of these activities during Choice Time. At the end of Session 9, the class discussion focuses on addition combinations that are related to combinations of doubles. These are referred to as doubles plus or minus 1.

Session 10: Today's Number and the Magic Pot As a class, students generate number sentences for Today's Number. They try to use combinations of 10, doubles, or doubles plus 1 in their expressions. During the second part of the session, students solve a Magic Pot problem that can be used for assessment.

Session 11: Counting Pockets Students collect data about the number of pockets worn by people in the class. As a group, they post the data on the board and look for familiar combinations of numbers, such as combinations of 10's, to help them add the number of pockets worn by the class. During the second half of the session, students complete Choice Time activities.

Mathematical Emphasis

- Developing familiarity of 10 as an important number in our number system
- Becoming familiar with number combinations of 10 and doubles
- Communicating about mathematical thinking through written and spoken language
- Developing strategies for adding two or more numbers
- Exploring the calculator as a tool for problem solving

What to Plan Ahead of Time

Materials

- Interlocking cubes: about 30 per student (All sessions)
- Chart paper (All sessions)
- Blank paper, 5½"-by-8½": 11 sheets per student (Session 1)
- Student math folders: 1 per student (Session 1)
- Number cards: 1 deck per pair (Sessions 2–3, and 8–9). If you do not have manufactured sets, make your own; see Other Preparation.
- Magic Pot items such as a dime, a box of markers, a bag containing 6–8 cubes (Sessions 4 and 5)
- Construction paper, white or light color, 12"-by-18": at least 1 sheet per student plus extras (Sessions 4 and 5)
- *Two of Everything* by Lily Toy Hong (A. Whitman, 1993) (Sessions 4 and 5, optional)
- Markers or crayons (Sessions 4 and 5)
- Overhead projector (Session 6, optional)
- Clipboards: 1 per student (Sessions 8 and 9, optional)
- Calculators: at least 1 per pair; if possible, 1 per student (Sessions 7–9)
- Large jar (Session 11)
- Resealable plastic bags or envelopes (to store card decks): 2 per student

Other Preparation

For Session 1

- Staple 11 sheets of 5½"-by-8½" paper together to make booklets, 1 per student.
- Duplicate Student Sheet 1, Weekly Log, 1 per student. At this time, you may wish to duplicate a supply to last for the entire unit and distribute the sheets as needed. Prepare a math folder for each student, if this was not done for a previous unit.
- Familiarize yourself with the classroom routines described at the end of this unit.

For Sessions 2 and 3

- If you do not have manufactured number cards from the grade 2 *Investigations* materials kit, use Student Sheets 2–5 to make one deck of number cards per pair of students for classwork. The classroom decks will last longer if duplicated on oaktag. These sets can be cut apart and stored in envelopes or plastic bags. Also duplicate enough to provide one deck per student for homework. Each deck should contain 4 of each number 0–9 and 4 wild cards. See the Materials lists with speciific activities to determine whether wild cards should be included with decks each time.
- Duplicate Student Sheet 6, Tens Go Fish, and Student Sheet 7, Turn Over 10, 1 each per student, plus 6–7 of each for the classroom. Familiarize yourself with these games.

What to Plan Ahead of Time (*continued*)

- Duplicate the Family letter (p. 152), 1 per family. Remember to sign and date the letter before copying it.

For Session 6

- Duplicate Student Sheet 8, Number Strings, and Student Sheet 9, More Number Strings, 1 each per student.

For Sessions 8 and 9

- Duplicate Student Sheet 10, Close to 20 Score Sheet, at least 2 per student, and Student Sheet 11, Close to 20 (Directions), 1 per student plus 6–7 for the classroom.

- Duplicate Student Sheets 12 and 13, Beat the Calculator Cards, 1 set per pair. Cut apart the cards to make decks. Decks will last longer if duplicated on oaktag. Store decks in envelopes or plastic bags.

For Session 10

- Duplicate Student Sheet 14, Our Class and the Magic Pot, 1 per student.

For Session 11

- Prepare a class list of names on chart paper or the chalkboard.

- Use the Pocket Data Chart from the previous unit. If you are not using the full-year *Investigations* curriculum, see About Classroom Routines, How Many Pockets? (p. 144) for information.

Combinations of 10

What Happens

Students work with combinations of 10. They list ways of making 10 with two and three numbers. They make a *Book of 10*, listing all of the possible combinations of 10 using two to ten addends. Their work focuses on:

■ exploring combinations of 10

Start-Up

Today's Number   Today's Number is one of three routines that are built into the grade 2 *Investigations* curriculum. Routines provide students regular practice in important mathematical ideas such as number combinations, counting and estimating data, and concepts of time. For Today's Number, which is done daily (or most days), students write number sentences that equal the number of days they have been in school. The complete description of Today's Number (p. 141) offers suggestions for establishing this routine, and some variations.

If you are doing the full-year grade 2 *Investigations* curriculum, you will have already started a 200 chart and a counting strip during the unit *Mathematical Thinking at Grade 2*. Write the next number on the 200 chart and add the next number card to the counting strip. As a class, brainstorm ways to express the number.

If you are teaching an *Investigations* unit for the first time, here are a few options for incorporating Today's Number as a routine:

■ **Begin with 1**   Begin a counting line that does not correspond to the school day number. Each day add a number to the strip and use this number as Today's Number.

■ **Use the Calendar Date**   If today is the sixteenth day of the month, use 16 as Today's Number.

Once Today's Number has been established, ask students to think about different ways to write the number. Post a piece of chart paper to record their suggestions. You might want to offer ideas to help students get started. If Today's Number is 45, you might suggest forty-five, 40 + 5, or 20 + 25.

Ask students to think about other ways to make Today's Number. List their suggestions on the chart paper. Occasionally as students offer suggestions, ask the group if they agree with the statements. In this way, students have the opportunity to confirm an idea that they might have had or to respond to an incorrect suggestion.

Materials

■ Interlocking cubes (10 per student)

■ Prepared booklets of 5½"-by-8½" paper (1 per student)

■ Chart paper

■ Student Sheet 1 (1 per student)

■ Student math folders (1 per student)

As students grow more accustomed to this routine, they will begin to see patterns in the combinations, have favorite kinds of number sentences, or use more complicated types of expressions.

Activity

Ways to Make 10

Distribute paper and interlocking cubes. Each student should have 10 cubes of the same color.

Suppose I want to divide 10 cubes into two groups. What is one way I could do that?

Take students' suggestions. For example:

Yes, I could have 4 cubes in one group and 6 cubes in another group. 4 and 6 make 10.

Now, working with partners, I would like you to investigate all the possible ways to divide 10 cubes into two groups. Make a list of all the combinations you find. When you think that you have them all, talk with your partner about why you are sure that you have found all the possible combinations of 10 using two numbers.

As students are working, observe their strategies for solving this problem. Do they use cubes? Do they list combinations in an organized way? Do they use one combination to derive another? Do some students just seem to *know* the combinations of 10 using two numbers?

When pairs of students finish, remind them to discuss how they know they have all the possible combinations. If some pairs finish early, suggest that pairs get together and explain their strategy for knowing that they have found all the possible combinations.

Call the class back together and ask students to share their combinations of 10 using two numbers. List students ideas on chart paper. If students offer suggestions that have already been mentioned, consider putting a check next to that combination, acknowledging each student's contribution.

Some students might suggest 10 + 0 as a possible combination. Acknowledge that this does make 10 and explain that in some of the activities they will be doing they will be using the combination 10 + 0. But in this activity each group needs to have at least one cube in it. You might list 10 + 0 at the bottom of the chart paper or off to the side.

There may be some discussion about whether combinations such as 4 + 6 and 6 + 4 are the same or different. Encourage students to share their thinking about this idea and accept both perspectives for this activity. Record these combinations as pairs, next to each other.

COMBINATIONS OF 10

√√4 + 6	6 + 4
5 + 5	
√7 + 3	3 + 7
9 + 1	1 + 9
√√8 + 2	2 + 8

When all combinations of 10 cubes have been listed, ask students how they know all the possible combinations have been found. As students offer their ideas, record their strategies on chart paper. In this way you model for students how their strategies and ideas can be recorded in words and shared with others.

There will probably be a variety of strategies for finding all the possible combinations of 10 using two numbers. Some students might make an organized list, others might show it using cubes, and others will use a guess-and-check method. The **Dialogue Box**, Finding Combinations of 10 (p. 27), provides some examples of how second graders approach the problem.

Phoebe

1 5+5=10
2 4+6=10
3 6+4=10
4 7+3=10
5 3+7=10
6 9+1=10
7 1+9=10
8 8+2=10
9 2+8=10

We are trying to get the number 10.

We got nine ways to get the number 10.

10 is my favirte number.

1 0

Book of 10

Referring to the list of combinations of 10 that the class just generated, pose the following problem:

We just made a list of combinations of 10 using two numbers. Suppose we wanted to use three numbers to make 10. For example, I have 10 cubes and I want to put them into three groups. How many cubes might be in each group? Remember, each group should have at least 1 cube in it, so zero is not a possibility. Think about this problem for a few minutes.

Encourage students to use their cubes to solve this problem. When most students are ready, ask a few to share their solutions. Since students will be adding these combinations to their *Book of 10*, list only two to three solutions.

There are many different ways to make 10, depending on how many different groups you use. We talked about combinations of 10 using two numbers, and you just began to think about three numbers. Over the next few days you will have a chance to explore ways to make 10 using many different combinations of numbers. Each of you will have a book to keep track of all the ways you come up with.

Distribute a booklet to each student and explain that they will keep track of the combinations of 10 by recording all the combinations they find in their own book. Have students title their book *Book of 10*, then write *1 Number* at the top of the first page, *2 Numbers* at the top of the second page, and so on, up to *10 Numbers* at the top of the last page.

Ask students to look at the page they labeled *2 Numbers*. Point out that on this page they should list all the possible ways to make 10 using two numbers. Explain that on the page labeled *3 Numbers*, they can record all the solutions they just found for the three numbers.

For the remainder of this session, students begin to add combinations to their *Book of 10*. They can work alone or with partners, but each student should make his or her own book.

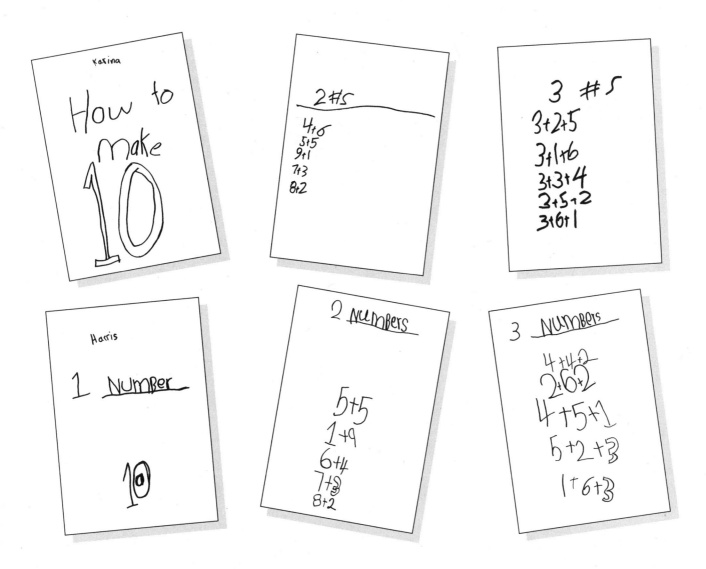

If you are using the full-year *Investigations* curriculum, students will be familiar with math folders and Weekly Logs. If this curriculum is new to students, tell them about one way they will keep track of the math work they do.

Mathematicians show how they think about and solve problems by talking about their work, drawing pictures, building models, and explaining their work in writing so that they can share their ideas with other people. Your math folder will be a place to collect the writing and drawing that you do in math class.

Distribute a math folder to students and have them label it with their names.

Introducing Math Folders and Weekly Logs

Your math folder is a place to keep track of what you do each day in math class. Sometimes there will be more than one activity to choose from, and at other times, like today, everyone in the class will do the same thing. Each day you will record what you did on this Weekly Log.

Students should put their *Book of 10* and the paper they used to list their combinations of 10 into their folders.

Distribute Student Sheet 1, Weekly Log, and ask students to write their name at the top of the page. Point out that there are spaces for each day of the week and ask them to write today's date on the line after the appropriate day. If you are doing the activity Today's Number, students can write the number in the box beside the date.

Ask students for suggestions about what to call today's activity. Titles for choices and whole-class activities should be short to encourage all students to record what they do each day. List their ideas on the board and have students choose one title to write in the space below the date.

❖ **Tip for the Linguistically Diverse Classroom** Encourage students who are not writing comfortably in English to use drawings to record in their Weekly Log. If students demonstrate some proficiency in writing, suggest that they record a few words with their drawings. Students can also record a sample problem representative of each day's work.

Weekly Logs can be stapled to the front of the folders (each new week on the top so prior logs can be viewed by lifting up the sheets).

During the unit, or throughout the year, you might use the math folders and Weekly Logs in a number of ways:

■ to keep track of what kinds of activities students choose to do and how frequently they choose them

■ to review with students, individually or as a group, the work they've accomplished

■ to share student work with families either by sending logs and/or folders home periodically for students to share or during student/family/teacher conferences

Finding Combinations of 10

In this classroom, the discussion during the activity Ways to Make 10 (p. 22) focuses on whether 3 + 7 is the same thing as 7 + 3. Some students see these as the same and label them "reversibles" or "pairs" of combinations, while others see the two combinations as different. The teacher's question about finding all the possible combinations of 10 using two numbers not only extends the discussion but also brings out some important observations about patterns in addition.

Who can tell me a way to make 10 using two numbers?

Paul: 8 + 2.

Carla: 9 + 1.

Phoebe: 2 + 8.

Lila: We already have that one.

Tim: I have two things to say. Lila is right. 8 + 2 and 2 + 8 are reversibles. And mine is 5 + 5. That's a double.

Let's talk about that idea of reversibles in one minute. Are there any other combinations that you can think of?

Angel: 6 + 4 and 3 + 7.

Ayaz: 7 + 3.

Lila: We already have that one too.

Someone raised the question about 3 + 7 and 7 + 3. He called them reversibles. What do you think about that?

Karina: 1 + 9 is the same as 9 + 1. It's just turned around.

Lila: It's the same amount in each pile, it's just that you put one number first. If you add 3 to 7 it's the same as adding 7 to 3. Look, I'll show you with the cubes. [*Lila puts a tower of 7 cubes and a tower of 3 cubes on the rug.*] See? This is 7 + 3. And this [*she moves the tower of 7 on the other side of the tower of three*] is 3 + 7. It's the same amount. So they are the same thing.

Helena: Every one of these combinations of 10 is part of a pair; 3 + 7 and 7 + 3, 2 + 8 and 8 + 2.

Ayaz: I say that they are different. See, you could have 3 red cubes and 7 blue cubes and that would be different from having 7 red cubes and 3 blue cubes.

Tim: Well, the colors are different, but the total amount of cubes you end up with is still the same. 10 is 10. I think they are reversibles because you could just flip them around and they would be the same. Or like if you read it in reverse, 3 + 7 would be 7 + 3.

You've all brought up some interesting ideas. Tim's idea of calling these addition combinations reversibles and Helena's observation that each combination is part of a pair are things for us to think about. I have another question for you. How did you figure out whether you found all the possible combinations of 10 using two numbers?

Linda: I said 9 + 1 and then just kept going lower and lower, like, 8 + 2, 7 + 3, 6 + 4 and, when I couldn't go any lower that was all of them.

Tory: I used cubes and I tried all the different ways to break them up and put them in two groups.

Did you have any way of keeping track?

Tory: No.

Tim: Our group did. We wrote 1 through 5 going down the paper and then I put plus signs in the middle and wrote 5 through 9 going up then we went back and checked. It's like a pattern, as one number goes down the other number goes up.

Continued on next page

continued

The teacher records Tim's method on the board to confirm his description.

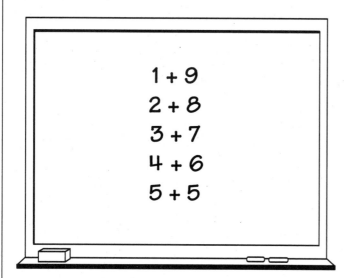

$$1 + 9$$
$$2 + 8$$
$$3 + 7$$
$$4 + 6$$
$$5 + 5$$

Helena: If you include all the pairs then I think there's only nine ways to make 10 with two numbers. You would take all the ways that Tim just said and double that because of the pairs. That would give you ten ways but 5 + 5 doesn't really have a pair so subtract 1 from 10 and that's 9.

So are you certain that you have all the ways to make 10 ?

Angel: Well, I got nine ways counting the pairs but I'm not absolutely certain that's all of them.

Linda: I'm certain because if you look at the pattern of going 1 down and 1 up then there aren't any numbers in between.

Tory: And besides, I can't think of any more combinations.

Card Games

What Happens

Students are introduced to two card games, Tens Go Fish and Turn Over 10. These, along with *Book of 10*, are choices for the next two sessions. Their work focuses on:

■ making 10 with two or more addends

Start-Up

Today's Number   Sometime during the school day, students brainstorm ways to express the number of days they have been in school. Add a card to the class counting strip and fill in another number on the blank 200 chart.

Materials

■ Number cards, see specific activities as to whether to include wild cards (1 deck per pair)

■ Student Sheets 6 and 7 (1 each per student for homework, plus 6–7 for the classroom)

■ Family letter (1 per family)

■ Interlocking cubes (about 10 per student)

Tens Go Fish and Turn Over 10 are two card games that reinforce combinations of 10 with two or more addends. If you are doing the full-year grade 2 *Investigations* curriculum, students will be familiar with these games from the unit *Mathematical Thinking at Grade 2*. Depending on the needs of your class, students may already be very comfortable and familiar with the combinations of 10 and you may want to skip these sessions and proceed to Session 4. You may, though, have some students who would benefit from playing these card games again. If this is the case, the games can easily be integrated into the Choice Time that occurs later in this unit.

If these games are new to students, try to teach both of them during the first half of Session 2. Students will work on these two activities and their *Book of 10* during the remainder of Session 2 and all of Session 3.

Note: If you have duplicated the number cards on paper instead of oaktag, have students make a "card holder" so they won't be able to see through the cards. Fold a sheet of 8½"-by-11"paper in half the long way and place it so that the fold is on the bottom. Fold the bottom edge up approximately 1 inch. Staple each side and slide cards into the "pocket."

Tens Go Fish and Turn Over 10

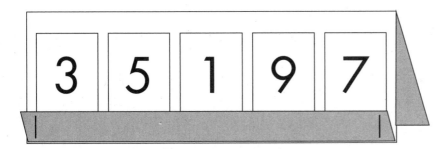

Tens Go Fish If students are familiar with the game Go Fish, they will need just a brief introduction to this game. If this is new to students, you might want to play a demonstration game in front of the class. Explain that each pair will need a deck of number cards. For this game, students should remove the wild cards and set them aside.

Today we will play a number card game. This game is called Tens Go Fish, and the object is to make matches or pairs of cards that add to 10. For example, which card would you pair with 7 to make 10? Each player is dealt five cards. You might be dealt two cards that equal 10, but most of the time you will have to ask someone else if he or she has a certain card. If I were holding these cards: 0, 2, 4, 5, and 7 [*write these numbers on the board*], **would I have a pair that totals 10?**

At my turn, I can ask another player for a card. For example, I would like to use my 4 card to make a 10. What card would I need to add to the 4 to make 10? So if I was playing with Trini I might ask, "Trini, do you have a 6?" If Trini has a 6, she gives it to me. I put the 6 and the 4 down as a pair, take a card from the deck, and my turn is over. If Trini does not have a 6, she says "Go Fish." I take a card from the deck and then my turn is over—even if I picked the card I asked for.

Explain that the game continues with each player trying to make combinations of 10. If at any time a player is holding two cards that total 10, he or she can put them down without drawing another card. The game is over when there are no more cards. At the end of the game, each player should make a list of the pairs of cards he or she has. Players label this list *Tens Go Fish*, write the date on the top, and put it in their math folders. Model this procedure for students. This is another way that you and students can keep track of their activity during Choice Time.

Depending on how much time is left in this session, you may want to have everyone in your class play a game of Tens Go Fish and introduce Turn Over 10 in the next session.

Note: You may want to suggest that students use cubes to help them find combinations of 10.

Turn Over 10 This is another card game that involves making combinations of 10. While Tens Go Fish uses combinations of two addends, in Turn Over 10, students make combinations of 10 using two or more addends.

The easiest way to introduce this game is by assembling students in a circle on the floor and playing a demonstration game in the middle of the circle.

This game is called Turn Over 10. You will be using all of the cards in the deck including the wild cards. A wild card is a special card because whoever turns it over can decide how much it is worth, depending on what is needed. To begin this game, put out the cards facedown in four rows with five cards in each row.

Arrange the cards. Pause and ask students to figure out how many cards there are. Notice how students figure out the total. Do they count each card by 1's? Do they count by 5's? Do they add 2 rows of 5 and then add 10 plus 10? Ask a few students to share their strategies.

Explain to students that the object is to turn over cards that total 10. Ask one student to play with you and demonstrate a few rounds in front of the class.

Turn over two cards. As you play, involve students in your turn.

I just turned over a 3 and a 4. What card could I turn over next to make exactly 10? What card could I turn over and make less than 10? What if I turn over a wild card, what should I make it into?

Demonstrate how cards from the deck are used to replace the cards that are taken when a total of 10 is made. You probably will not need to complete an entire demonstration game. Make sure, however, that students understand that the game is over when no more combinations of 10 can be made. At the end of the game, each player should record on a piece of paper his or her combinations in a way similar to Tens Go Fish. They write the name of the game, the date, and the combinations of cards they made.

Introducing Choice Time

Explain to students that in math class they often will be working on Choice Time activities. This means they will have to choose which of several different activities they will work on.

Choice Time is a format that recurs throughout the *Investigations* curriculum. See the **Teacher Note**, About Choice Time (p. 35), for information about how to set up Choice Time, including how students might use their Weekly Logs to keep track of their work.

Explain how students are to work during Choice Time. Each day, they select one or two of the activities they want to participate in. Students can select the same activity more than once, but they should not do the same activity each day. Decide whether you want students to do every activity or just some of them.

List these choices on the chalkboard. You may want to have a few copies of the game directions (Student Sheets 6 and 7) available for students to refer to as they play. For the *Book of 10* activity, students continue adding combinations to their *Book of 10* that they began in Session 1.

1. Tens Go Fish
2. Turn Over 10
3. Book of 10

Choice 1: Tens Go Fish

Materials: Deck of number cards 0–10 (without wild cards) for each group of 2–3 players; interlocking cubes available as counters

Each player is dealt five cards. Players take turns asking each other for cards that when combined with a card in their hand will total 10. If a player makes a combination of 10, those cards are placed on the table, a new card is drawn from the deck, and the player's turn is over. If a player cannot make a combination of 10, a card is drawn from the deck. If the player draws the card he or she requested, the player can put down the matching pair; then the turn is over. The *game* is over when there are no more cards left in the deck. At the end of the game each player makes a list of the combinations of 10 that he or she collected.

Choice 2: Turn Over 10

Materials: Deck of number cards 0–10 (with wild cards) for each group of 2–3 players; interlocking cubes available as counters

Special Education 406

Show and Tell

Name Laurie Labato-DiTomasso   **Date** September 15, 1997

Description Pocket Day

This activity provides children with meaningful practice of addition, counting and estimation skills. Students use the number of pockets they are wearing on Pocket Day to provide the data for the lesson. Additional materials needed are chalk board or chart paper, counters (interlocking cubes, pennies, tongue depressors with colored dots in the sets needed) pencils and paper for each student and optional estimation jar with rubber band placed around it.

How Used

Step 1. Students estimate how many pockets the class is wearing today. Record the estimates of the group.

Step 2. Each student counts the number of pockets they are wearing.

Step 3. Starting with children with no pockets and working your way up each set the students are called in front of the class to hold a symbol for the number of pockets they are wearing. The entire class count the number of pockets in each set. Example : All students with one pocket are called to the front of the class. They each take one counter. The teacher asks the class how many pockets does this group of children have on today. Student give their responses and the teacher asks how they achieved their answers. The answer is verified by the entire class and recorded along with a number sentence that symbolizes the total. Students may keep a running total of the data at their seat. If using interlocking cubes, they may be put into the class estimation jar. The next group is called forward and they go through the same process. The sums of both groups are totaled. The children will begin to see that certain ways of counting are quicker than others and learn various ways to create number sentences.

Step 4. After all data has been collected compare beginning estimate with the actual number of pockets. If using an Estimation Jar place the elastic band at the level that the markers have reached in the jar. During your next pocket day you can now compare the results of both days.

Step 5. Optional: Students may work in small groups after the whole group lesson verifying the number of cubes in the jar .

Cost

tongue depressors $3.00, colored dots $2.79, estimation jar can be gotten free from recycled items, cubes used were supplied by school, chart paper $.50

Arrange the cards facedown in four rows of five. The remaining cards are placed facedown in a pile. Players take turns trying to turn over cards that total 10. If the total is more than 10, the turn is over and the cards are turned facedown again. If the total is exactly 10, players take the cards, replace the missing cards with cards from the deck, and continue their turn. The game is over when no more 10's can be made. At the end of the game each player records the combinations of 10 he or she found during the game.

Choice 3: *Book of 10*

Materials: *Book of 10* from previous session, 10 cubes per student

Students continue adding more combinations to their *Book of 10* that they began in Session 1. They can work in pairs, but each student should make his or her own book. Cubes should be available for students to use as they look for combinations of 10.

Observing the Students

During this first Choice Time, observe students at work.

Do students try each choice, or do they stay with a familiar one? If, after a short time with one activity, students say they're done, ask them to tell you about what they have done and encourage them to investigate further. Do students work alone or with partners? Do they share what they have done with others and observe what others are doing? Do they talk to themselves or others about what they are doing?

Below are some specific suggestions of what you might observe as students work on Choice Time activities. These activities focus on combinations of numbers and on counting.

Tens Go Fish

- Do students easily identify pairs of 10?
- How do they decide which card to ask for?
- Do they use cubes?
- Do they count on from the number they have?
- Do they just "know" the pairs of numbers that make 10?
- Are students able to keep track of the cards other players have asked for?

Turn Over 10

- How do students keep track of the total they have turned over?
- Are students able to predict which card they need to make a total of 10?
- Are students able to keep track of the position of numbers or do they randomly turn over cards?
- How do students use wild cards to complete their 10 or to extend their turn?

Book of 10

■ How do students generate combinations?

■ Do they use patterns?

■ Do they work in an organized way or do they generate combinations by trial and error?

■ Do they use cubes to help them figure out combinations?

■ Do they add up to get to 10 or do they work backward from 10?

■ How do students record their combinations?

If students are stuck and are able to list only one or two ways to make 10 using four numbers in their *Book of 10*, you may suggest that they investigate what happens when one of those four numbers is written in a different way. **I see you found that 5 + 3 + 1 + 1 equals 10. If you kept the 3 + 1 + 1, is there another way to make 5 so that the total comes out to be 10?**

Near the End of the Session   Five or ten minutes before the end of each Choice Time session, have students stop working, put away the materials they have been working with, and clean up their work area.

When cleanup is complete, students should record on their Weekly Logs what they worked on during Choice Time. Suggest that they use the list of Choice Time activities that you posted as one reference for writing about what they did during Choice Time.

Whenever possible, either at the beginning or end of Choice Time, have students share the work they have been doing. This often sparks interest in an activity. Some days you might ask two or three students to share with the class the work they have been doing. On other days you might ask a question that came up during Choice Time so that others might respond to it. Sometimes you might want students to explain how they thought about or solved a particular problem.

 **Homework**

Sessions 2 and 3 Follow-Up

At the end of Session 3, assign students to play one of the card games with someone at home. Each student will need to take home number cards that they can cut up at home and directions for the game they have chosen to teach their family, Tens Go Fish (Directions) or Turn Over 10 (Directions). You also may want to provide an envelope for them to keep their deck of number cards in. Suggest that students keep these cards in a special place because they will be playing these games and other card games for homework throughout the unit.

Send home the Family letter with this homework.

About Choice Time

Choice Time is an opportunity for students to work on a variety of activities that focus on similar mathematical content. Choice Time activities are found in most units of the grade 2 *Investigations* curriculum. These generally alternate with whole-class activities in which students work individually or in pairs on one or two problems. Each format offers somewhat different classroom experiences. Both are important for students to be engaged in.

In Choice Time the activities are not sequential; as students move among them, they continually revisit some of the important concepts and ideas they are learning. Many Choice Time activities are designed with the intent that students will work on them more than once. By playing a game a second or third time or solving similar problems, students are able to refine strategies, see a variety of approaches, and bring new knowledge to familiar experiences.

You may want to limit the number of students who work on a Choice Time activity at one time. Often when a new choice is introduced, many students want to do it first. Assure them that they will be able to try each choice. In many cases, the quantity of materials available limits the number of students who can do an activity at any one time. Even if this is not the case, set guidelines about the number of students who work on each choice. This gives students the opportunity to work in smaller groups and to make decisions about what they want and need to do. It also provides a chance to return and do some choices more than once.

Initially you may need to help students plan what they do. Rather than organizing them into groups and circulating among the groups every 15 minutes, support students in making decisions about the choices they do. Making choices, planning their time, and taking responsibility for their own learning are important aspects of a student's school experience. If some students return to the same activity over and over again without trying others, suggest that they make a different first

choice and then choose the favorite activity as a second choice.

How to Set Up Choices

Some teachers prefer to have choices set up at centers or stations around the room. At each center, students find the materials needed to complete the activity. Other teachers prefer to store materials in a central location and have students bring materials to their desks or tables. In either case, materials should be readily accessible to students, and students should be expected to take responsibility for cleaning up and returning materials to their appropriate storage locations. Giving students a "5 minutes until clean up" warning before the end of an activity session allows them to finish what they are working on and prepare for the upcoming transition.

You may find that you need to experiment with a few different structures before finding a setup that works best for you and your students.

The Role of the Student

Establish clear guidelines when you introduce Choice Time activities. Discuss students' responsibilities:

- Try every choice at least once.
- Work with a partner or alone. (Some activities require that students work in pairs, while others can be done either alone or with partners.)
- Keep track, on paper, of the choices you have worked on.
- Keep all your work in your math folder.
- Ask questions of other students when you don't understand or feel stuck. (Some teachers establish the rule, "Ask two other students before me," requiring students to check with two peers before coming to the teacher for help.)

Students can use their Weekly Logs to keep track of their work. As students finish a choice, they write it on their log and place any work they

Continued on next page

have done in their folder. Some teachers list the choices for sessions on a chart, the board, or the overhead projector to help students keep track of what they need to do.

In any classroom there will be a range of how much work students complete. Some choices include extensions and additional problems for students to do when they have completed their required work. Encourage students to return to choices they have done before, do another problem or two from the choice, or play a game again.

At the end of a Choice Time session, spend a few minutes discussing with students what went smoothly, what sorts of issues arose and how they were resolved, and what students enjoyed or found difficult. Encourage students to be involved in the process of finding solutions to problems that come up in the classroom. In doing so, they take some responsibility for their own behavior and become involved with establishing classroom policies. You may also want to make the choices available at other times during the day.

The Role of the Teacher

Choice Time provides you with the opportunity to observe and listen to students while they work. At times, you may want to meet with individual students, pairs, or small groups who need help or whom you haven't had a chance to observe before, or to do individual assessments. Recording your observations of students will help you keep track of how they are interacting with materials and solving problems. The **Teacher Note,** Keeping Track of Students' Work p. (37), offers some strategies for recording and using observations of students.

During the initial weeks of Choice Time most of your time probably will be spent circulating around the classroom helping students get settled into activities and monitoring the overall classroom management. Once routines are familiar and well established, students will become more independent and responsible for their work. This will allow you to spend more concentrated periods of time observing the class as a whole or working with individuals and small groups.

Keeping Track of Students' Work

Throughout the *Investigations* curriculum, there are numerous opportunities to observe students as they work. Teacher observations are an important part of ongoing assessment. While individual observations are snapshots of a student's experience with a single activity, when considered over time they can provide an informative and detailed picture. These observations can be useful in documenting and assessing a student's growth. They offer important sources of information when preparing for family conferences or writing student reports.

Your observations of students will vary throughout the year. At times you may be interested in particular strategies that students are developing to solve problems. Or you might want to observe how students use or do not use materials to help them solve problems. At other times you may be interested in noting the strategy that a student uses when playing a game during Choice Time. Class discussions also provide many opportunities to take note of students' ideas and thinking.

You will probably find it necessary to develop some sort of system to record and keep track of your observations of students. While a few ideas and suggestions are offered here, the most important aspect of developing a tracking system is finding one that works for you. Too often keeping observation notes on a class of 28 students can become overwhelming and time-consuming.

A class list of names is a convenient way of jotting down observations of students. Since the space is limited, it is not possible to write lengthy notes; however, when kept over time, these short observations provide important information.

Stick-on address labels can be kept on clipboards around the room. Notes can be taken on individual students and then these labels can be peeled off and stuck into a file that has been set up for each student.

Alternatively, jotting down brief notes at the end of each week may work well for you. Some teachers find that this is a useful way of reflecting on the class as a whole, on the curriculum, and on individual students. Planning for the next weeks' activities often develops from these weekly reflections.

In addition to your own notes on students, each student will be keeping a folder of work for any given unit. This work and the daily entries on the Weekly Logs also can document a student's experience. Together they can help you keep track of the students in your classroom, assess their growth over time, and communicate this information to others. At the end of each unit there is a list of things you might choose to keep in students' folders.

Doubles

Materials

- *Two of Everything* by Lily Toy Hong (optional)
- Magic Pot items such as a dime, a box of markers, a bag containing 6–8 cubes
- Chart paper
- Interlocking cubes (about 30 per student)
- Construction paper, white or light colored (at least 1 sheet per student plus extras)
- Markers or crayons

What Happens

The book *Two of Everything* introduces students to the concept of doubling. They generate a list of items and then figure out how many they would have if they were doubled. As a class they make a book of doubles riddles. Their work focuses on:

- doubling numbers
- recording strategies for doubling numbers

Start-Up

Today's Number Suggest that students use combinations of 10 in their number sentences. For example, if the number they are working on is 34 and one number sentence is $10 + 10 + 10 + 4$, ask students if there is another way of making 34, such as $6 + 4 + 6 + 4 + 6 + 4 + 4$. Add a card to the class counting strip and fill in another number on the blank 200 chart.

Activity

Two of Everything

Two of Everything is a story about a man and a woman who find a magic pot. By accident they discover that the pot doubles whatever is put into it. When the woman drops her hairpin into the pot, to her surprise, two hairpins come out of the pot. When she puts her purse containing five gold coins into the pot, she pulls out two identical purses each containing five gold coins! The man and woman test a variety of items, each time getting back double what they put in. Then the man and his wife fall into the pot and even they are doubled—creating two identical men and two identical women. The story ends with each couple living side by side, each with identical houses containing identical possessions.

If you have the book *Two of Everything*, read it to your class. If not, retell the story in your own way. The most important idea to communicate is that the magic pot doubles whatever is put into it. As you read the book or tell the story, pause each time something is put into the pot and ask students to predict what will come out and how many things there will be.

In this story the pot had magic powers. It made two of everything that was put into it. Suppose you had a chance to put something into the magic pot. What would you put in? Why?

Suppose I put in a box that has 8 markers. What would come out of the magic pot?

Some students will say two boxes of markers. Accept this answer and ask how many markers there would be. Take students' suggestions. For example: Yes, there would be 2 boxes of markers [*write 1 + 1 = 2 on the board or chart paper*]. How many markers would there be? How could I record that using numbers? [*Leave some space under 1 + 1 = 2 and record 8 + 8 = 16 near the bottom.*]

What if I put in a pair of shoes? How many shoes would I get out?

Again ask students to suggest how to write what happened using a number sentence. List 2 + 2 = 4 under 1 + 1 = 2.

What if I put in the three bears? Who can give me a number sentence that describes what would happen? [*Pause as students respond.*] One more. If I put these four cubes in the pot, what would come out?

I'd like you to look at these number sentences and talk with someone about what you notice.

1 + 1 = 2 (box of markers)
2 + 2 = 4 (pair of shoes)
3 + 3 = 6 (bears)
4 + 4 = 8 (cubes)

8 + 8 = 16 (markers)

After a few minutes ask students to share their observations. Record their ideas on the board or on chart paper.

Some students might notice that the totals go up by 2. Extend their thinking by asking if anyone can explain why this happens. If no one has an idea, leave the question as something the group will think about.

Book of Magic Pot Riddles

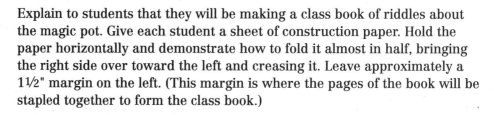

Explain to students that they will be making a class book of riddles about the magic pot. Give each student a sheet of construction paper. Hold the paper horizontally and demonstrate how to fold it almost in half, bringing the right side over toward the left and creasing it. Leave approximately a 1½" margin on the left. (This margin is where the pages of the book will be stapled together to form the class book.)

Show students how they will write their riddle on the front of the folded flap and how they will solve the problem by writing a number sentence and the sum inside the flap. Explain to students that they can try Magic Pot Riddles during Choice Time.

Each of you will make up a riddle for the riddle book. You can choose anything you would like to put into the magic pot. On the front flap of your paper you will write a sentence about what goes into the pot. I wrote this sentence on my paper: I collected one dozen eggs from the henhouse and put them in the magic pot. How many eggs will I get back?

On the inside of your paper you will write a number sentence that goes with your riddle, the answer to your riddle, and show how you would solve the problem.

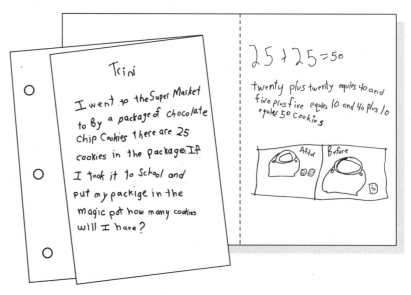

❖ **Tip for the Linguistically Diverse Classroom** Students with limited English proficiency can make a drawing of their riddle using symbols and numbers. Students can divide their paper into thirds. On the left side, they can draw what they are going to put into the pot. In the middle, they can draw the pot. And on the right side, have them put a question mark. On the back of the paper, students can write their number sentence.

Some students may need help choosing an appropriate number to work with in their riddle. Before students begin working, you might brainstorm possible ideas with them. Many second graders will be able to handle numbers larger than 10.

For the remainder of this session and half of the next, students work on writing magic pot riddles. During each session, have interlocking cubes available as counters. When students are finished, have them read their work to you. Give students feedback on the clarity of both their riddle and of their solution. Students' solutions to their riddle should include not only an answer but also an explanation either in words or numbers of how they solved the problem. When students are finished with one riddle, encourage them to write another to contribute to the class book.

Sharing Magic Pot Riddles

Leave approximately 30 minutes at the end of Session 3 for students to share their magic pot riddles. Before sharing with the whole class, organize students into pairs or triples and have them read their riddles to each other. Partners can try to solve each other's riddles. As students are doing this, circulate around the class and identify two or three interesting riddles to share during the class discussion. Try to identify students who have used different problem-solving strategies. This is one way to begin to engage students in strategies for combining two numbers.

Remind students to fill out their Weekly Logs at the end of each session.

Sessions 4 and 5 Follow-Up

🏠 Homework

After Session 4 or 5, students teach someone at home to play either Tens Go Fish or Turn Over 10. They will need directions for whichever game they have not yet played at home (Student Sheet 6 or 7).

Number Strings

What Happens

Students discuss strategies for adding "strings" of numbers. They use doubles (3 + 3, 4 + 4) and combinations of 10 (3 + 7, 4 + 6) to solve addition problems. They record their solutions and then discuss their strategies with partners. Their work focuses on:

- using doubles and combinations of 10 to solve related number combinations

Start-Up

Today's Number   Suggest that students use doubles in their number sentences. For example, if the number they are working on is 36, possible combinations include: 18 + 18, 9 + 9 + 9 + 9, or 10 + 10 + 8 + 8. Add a card to the class counting strip and fill in another number on the blank 200 chart.

Playing Card Games at Home   At the beginning of class briefly check in with students about their experiences teaching Tens Go Fish or Turn Over 10 to someone at home. Remind students that they should keep their playing cards and directions in a safe place at home because they will need them for future homework assignments.

Materials

- Student Sheet 8 (1 per student)
- Student Sheet 9 (1 per student for homework)
- Interlocking cubes (about 50 per pair)
- Overhead projector (optional)

Write 7 + 6 = _____ on the board or overhead.

Ask students to think of a story problem that could go along with this number sentence. Have students share a few of their story situations and note whether they reflect an addition or combining situation. Then ask students to think about how they would solve the problem.

Think about how you would solve 7 + 6. I'm interested in the problem (or situation) **you thought of, but now I'd like you to think about how you would solve the problem.**

Pause for students to think about the problem. As you call on students, remind them to explain how they added these two numbers. As students offer explanations, record their strategies on the board.

Activity

Strategies for Combining Two Numbers

Some students might count on from 6 or 7, using their fingers to keep track of the amount they are adding. This recording might look like:

7, 8, 9, 10, 11, 12, 13

Some students might count from 1 using counters or their fingers to show a group of 7 and a group of 6. You can record this strategy as:

1, 2, 3, 4, 5, 6, 7, 8, . . . 13

Other students might relate this combination to one of the doubles: "I know that 6 + 6 is 12 so 7 + 6 is 1 more" or "I know 7 + 7 is 14 so 7 + 6 is 1 less, 13." This could get recorded as:

6 + 6 = 12 and 12 + 1 = 13
or
7 + 7 = 14 and 14 - 1 = 13

As you record each strategy, ask if anyone else used the same strategy to solve the problem. In this way you acknowledge many students' thinking while moving the conversation along toward getting a variety of different strategies. Tell students that there are many different ways to solve the same problem. Then write, 4 + 5 = on the board.

Here's another problem for you to solve. When you are thinking about this problem, I'd like you to think about the strategies on the board and see if your strategy is the same as, or different from, any of those.

Again, give students a chance to think about the problem. Then, as students explain their strategy, record it on the board. As you do so you might add, **So you counted on from 4** or **You used a double (4 + 4 or 3 + 3) that you already knew and added (subtracted) 1. Is there anyone else who used a double plus 1?** In this way students will begin to see that there are ways of referring to their ideas in more general terms.

Here is one more problem we will try. How would you add 8 + 5 + 2?

There will probably be a variety of strategies. Some students will use the combination of 10 (8 + 2) and then add 5. Some might add 5 + 2 and then relate 8 + 7 to one of the doubles, while others may count the numbers in succession. As you gather strategies, emphasize that looking for combinations of numbers that you already know, such as the doubles and combinations of 10, are helpful when adding numbers. The **Teacher Note,** Strategies for Learning Addition Combinations (p. 46), provides other examples of similar strategies.

As you record strategies, use lines to indicate numbers that students thought of as a combination, such as:

8 + 5 + 2

10

Emphasize that students should look at the whole problem to see if they can identify numbers that go together, such as combinations of 10 or doubles or other familiar number combinations.

Distribute Student Sheet 8, Number Strings, to students. Make interlocking cubes available for students to use as counters. They should first solve these problems on their own and then check their work with partners. Students will record how they solved each problem using words, numbers, or pictures. Provide suggestions such as these:

When you write about how you solved a number string problem, you can use words to explain your strategy. For example, "I counted on from 4 and got to 7." Or you could write 4, 5, 6, 7 to show that you counted on. If you used a double or a combination of 10 to help you solve a problem, you might say, "I added 8 + 2 to get 10 and then added 5 more to get 15." Or you might just use numbers: "8 + 2 = 10 and 10 + 5 = 15." What is most important is that you communicate your strategy to someone else.

When you have finished, find a partner to compare strategies with. First explain to your partner how you solved a problem. Then your partner explains how he or she solved the same problem.

As students are working, try to get a sense of the strategies they are using.

■ How do students use materials? Do they make each number and then count all?

■ Are students looking at the whole problem or adding numbers in sequence?

■ Are there students who look for known or familiar combinations and work with these?

■ Are students using doubles to solve near-doubles problems?

■ How do students record their strategies?

As you are observing, ask students to explain their recording system to you. If students have missed a step it may become clear when they talk it out. If students have recorded incorrect totals, resist having them correct the problem. Most likely when they share their strategies with their partners the error will surface. You might want to make a note of students whom you want to observe during sharing time to see how they handle incorrect totals.

Session 6 Follow-Up

🏠 Homework

Students solve More Number String problems on Student Sheet 9. Remind them to record their strategy for each problem using words, pictures, or number sentences.

Teacher Note ⟩ *Strategies for Learning Addition Combinations*

To develop good computation strategies, students need to become fluent with the addition combinations from 0 + 0 to 10 + 10. These combinations are part of the repertoire of number knowledge that contributes to the rich interconnections among numbers that we call number sense. A great deal of emphasis has been put on learning these addition combinations in elementary school. While we agree that knowing these combinations is important, we want to stress two ideas:

■ Students learn these combinations best by using strategies, not simply by rote memorization. Relying on memory alone is not sufficient, as many of us know from our own schooling. If you forget—as we all do at times—you are left with nothing. If, on the other hand, your learning is based on understanding of numbers and their relationships, you have a way to rethink and restructure your knowledge when you don't remember something you thought you knew.

■ Knowing the number combinations should be judged by fluency in use, not necessarily by instantaneous recall. Through repeated use and familiarity, students will come to know most of the addition combinations quickly and a few by using some quick and comfortable numerical reasoning strategy. For example, when one of the *Investigations* authors thinks of 8 + 5, she doesn't automatically see the total as 13; rather, she sees the 5 broken apart into a 2 and a 3, the 2 combined with the 8 to make 10, then the 10 and the 3 combined to total 13. While this strategy takes quite a while to write down or to read, she "sees" this relationship almost instantaneously. As far as she is concerned, she "knows" this addition combination.

Using Strategies to Learn Addition Combinations

You will notice that we have avoided calling these addition combinations addition "facts."

Continued on next page

We avoid the term *facts* because it tends to elevate knowledge of these combinations above other mathematical knowledge—as if knowing these is the most important thing in mathematics. Developing fluency in using combinations is important, but there are other ideas that are just as critical for developing sound number sense.

There are 121 addition combinations from 0 + 0 to 10 + 10, and many of these are learned without difficulty. In this unit, it is assumed that students already know the combinations that involve + 0, + 1, and + 2. Most students solve these combinations quickly by counting on. They also need to recognize that, for example, adding 2 + 8 is the same as adding 8 + 2, so that they use the more efficient counting-on strategy (8, 9, 10), rather than beginning with 2 and counting up 8 more.

Excluding the + 0, + 1, and + 2 combinations, there are 36 combinations up to 10 + 10, if pairs of combinations (such as 7 + 3 and 3 + 7) are learned together. (It is important to note that not all second graders will see 3 + 7 and 7 + 3 as the same problem.) These 36 combinations can be grouped to help students learn good strategies for solving them easily. Here are some useful groupings:

- **The Doubles**—from 3 + 3 to 10 + 10. Students learn most of the doubles readily and can use the doubles they know to help with the harder doubles: "I know that 6 + 6 is 12, so 7 + 7 is 2 more, that's 14."

- **The Near Doubles**—3 + 4, 4 + 5, 5 + 6, 6 + 7, 7 + 8, 8 + 9, and 9 + 10. These are 1 away from the doubles. Students can use the doubles they know to learn these: "I know that 5 + 5 is 10, so 5 + 6 is 1 more" or "I know that 6 + 6 is 12, so 5 + 6 is 1 less."

- **Sums That Make 10**—3 + 7, 4 + 6, 5 + 5, 6 + 4, and 7 + 3. Students need many experiences building all the ways there are to make 10 with interlocking cubes until they recognize these combinations.

- **The 10+ Combinations**—from 10 + 3 to 10 + 10. Because these combinations follow a structural pattern, students learn them readily once they have built them repeatedly with cubes or counted them out on the 100 chart.

- **The 9+ Combinations**— from 9 + 3 to 9 + 10. Students can think of these combinations this way: To solve 9 + 6, take 1 from the 6 and add it to the 9 to make 10. The 5 that is left is added to the 10, 10 + 5 = 15. Or, if this were 10 + 6, the answer would be 16, but it's 1 less, so it's 15.

Some combinations belong to more than one group: 9 + 8 is a 9+ combination and also a near double. Students can use whatever strategies work best for them.

Only eight single-digit addition combinations do not fall into any of these categories above: 5 + 3, 6 + 3, 7 + 4, 7 + 5, 8 + 3, 8 + 4, 8 + 5, and 8 + 6. Many strategies can be used for each of these: relating them to doubles, to combinations that make 10, or to the 10+ combinations. Students should choose whichever strategies make the most sense to them.

In this unit students work with the strategies of combinations of 10, doubles, and near doubles. Work in later units will introduce students to other strategies. This work with addition combinations continues in the early part of third grade.

Exploring Calculators

What Happens

Students explore the calculator as a tool for mathematics. As they make up and solve problems on the calculator, you have a chance to see how familiar they are with this tool. In a class discussion, students share what they know about the calculator, including use of the operation keys and an awareness of the decimal point. As a class they solve number string problems. Their work focuses on:

- exploring the calculator as a mathematical tool
- using the calculator to solve computation problems

Materials

- Calculators (at least 1 per pair)

Start-Up

Today's Number   Sometime during the school day, students brainstorm ways to express the number of days they have been in school using three addends. Add a card to the class counting strip and fill in another number on the blank 200 chart.

More Number Strings   Spend the first 5 minutes of this session having pairs of students compare the solutions they used for solving the homework problems, More Number Strings.

Activity

Exploring Calculators

Students enjoy using what they perceive as adult tools, such as calculators. It is important that they have access to calculators as a tool for doing mathematics just as they have access to other materials such as 100 charts, pattern blocks, or interlocking cubes. As with any material it takes practice to use the calculator efficiently, accurately, and appropriately. Students will vary in their experiences and will need time to explore the calculator and find out what it can do. When calculators are used only occasionally, students become excited and distracted when they are permitted to use them. The focus is then on the calculators rather than on the mathematics.

To help students learn when it is appropriate to use calculators, they need to have easy access to them and use them frequently in their work. You will need at least one calculator per pair of students. Since this activity is an exploratory one, it is preferable to have a calculator for each student.

Mathematicians often use tools and materials to help them solve problems or to help them explain their thinking to others. This year you have been using tools like interlocking cubes, pattern blocks, pencil and paper, and Geoblocks to solve math problems. The calculator is another tool to use to solve problems.

Ask students to share how they have used a calculator or have seen a calculator being used. Make a list on the board or chart paper.

Distribute calculators and give students about 10 minutes to explore them. As you circulate around the classroom, get a sense of how familiar students are with the calculator:

- Are they familiar with the symbols on the keyboard?
- Can they do straightforward computation?
- Can they read the number on the screen?
- Do they know how to clear the screen and begin a new problem?
- Do they notice numbers with decimal points?

After students have explored the calculator, ask them to share one thing they notice about it or one thing they can do with it. As students share, listen for the opportunity to highlight these aspects of the calculator:

- how to turn the calculator on and off
- the numeral keys 0–9 and how to enter a two-digit number
- the operation keys for addition and subtraction
- the equals sign
- the screen display
- how to clear the screen display

Some students in your class probably will be familiar with one or many of these aspects. Encourage students to teach one another about them as they are mentioned by having a student talk through a problem that was solved on the calculator. As they do this, record the equation on the board, pointing out the use of the numeral key, the operation sign, and the equals sign.

Most of the calculations that second graders will be doing on calculators do not involve decimal numbers, but most likely they will bump into numbers with decimals when they are experimenting with the calculator. Bring it up if no one mentions it in the discussion.

Jeffrey and Juanita got this number [*write 12.46 on the board*] **when they were investigating the calculator. What do you think this dot means?**

Some students may have a variety of ideas about the decimal point on the calculator: "It's like a comma." "You use it in money." "It's like a period at the end of a sentence." "It's the extra part of the number." Some students may think that they have made a mistake or that the calculator made a mistake when they get a decimal number.

It is not critical that second graders be able to interpret and understand the meaning of the decimal point on the calculator. Instead, they should begin to think of it as showing a small part of a number that is less than 1. You might explain that Jeffrey and Juanita's number is 12 plus a little extra.

Activity

Problems to Solve on the Calculator

Distribute paper to students and ask them to make up some addition or subtraction problems to solve on the calculator. Students first record their problem on paper and then use the calculator to solve the problem and record the answer. If students are having difficulty generating their own problems to solve, suggest a few of the following: 6 + 9, 20 – 3, 15 + 11.

After you and your partner have each solved four or five problems and recorded them on your paper, trade papers and check each other's calculations. If you agree with your partner's calculation, put a check mark next to the number sentence. If you disagree, put a question mark next to the number sentence. When you are both finished, discuss the number sentences that you do not agree on and see if you can figure out a way to double-check your solutions.

Often when students are introduced to the calculator they are fascinated by the power of being able to make big numbers or solve complicated arithmetic problems. Students will approach this task in a variety of ways. Some may choose fairly straightforward problems to record and solve while others may choose more complicated problems involving larger numbers and possibly more than one operation. Some students will understand the problems they choose to record, while others may be able to understand the process of entering the numbers into the calculator but not necessarily assess whether the answer on the calculator is reasonable.

Students need time to explore this new tool and discover ways they can use it to help them solve problems (much the same as they have figured out ways of using cubes or blocks or pencil and paper to solve problems). Just as we want students to use these materials in ways that make sense to them, this is the goal for using the calculator. One way of fostering this is to check in with individuals or pairs of students as they are working, by asking questions like:

- Does the answer on the calculator make sense?
- Can you estimate about how much the answer to this problem will be?
- Are there any of these problems that you could do in your head? Could you show me?

Suggest that students choose one problem to solve first with the calculator and then with paper and pencil or mentally and see if they agree with the calculator.

As you circulate around the classroom, take note of certain problems that you would like students to share and discuss as a whole class. Choose a couple of problems that many students could solve mentally as well as with the calculator, such as one involving single digits and one adding a number in the 10's to a number in the 20's (17 + 21).

Sharing Problems

Leave 10 minutes at the end of the session for students to share some of the problems they solved on the calculator. Begin by asking a volunteer to share an addition problem. Write the problem on the board and have students use their calculators to compute the total.

Ask one student to talk through the process of solving the problem on the calculator. For example, for the problem 17 + 21, a student's explanation might be: "First I did the 17. I pushed 1 then 7. Then I pushed the plus sign and then I pushed the 2 and the 1, then the equals sign. I got the answer 38."

So, Ayaz got 38 as an answer to the problem 17 + 21. Did anyone get a different answer? [*Record those answers on the board.*]

Let's think about this problem. Turn to a person near you and explain what you think a reasonable answer is.

Ask a variety of students to share their strategies for thinking about the answer to this problem. Students will probably suggest different solution strategies. As students share their strategies, record their ideas on the chalkboard.

17, 27, 37, plus 1 more is 38
17, 18, 19, 20, . . . 38

10 + 20 = 30
7 + 1 = 8
30 + 8 = 38

Ask for another problem and follow the same procedure. Emphasize to students that when they are solving a problem with a calculator, with other materials, or in their heads, it is important to think each time about whether the answer makes sense. Often being able to solve a problem in more than one way or with a variety of materials is important.

At the end of this session, explain to students how the calculators will be stored. The **Teacher Note**, Using the Calculator Sensibly in the Classroom (p. 53), provides suggestions for calculator storage and usage.

Using the Calculator Sensibly in the Classroom

Why calculators in the elementary classroom? Many people are still opposed to the use of calculators in the elementary classroom. You may hear statements from parents or other educators such as: "Students should not use calculators until they know the facts and procedures" or "Students should use calculators to check only after they have solved a problem."

By keeping calculators out of the hands of young students in school, while they see many adults using them, we communicate clearly that doing mathematics in school is nothing like doing mathematics outside of school. Calculators are now a critical tool in our society. Increasingly sophisticated calculators are being developed and used in settings ranging from high school mathematics courses to jobs in science, business, and construction. Students need to learn how to use the calculator effectively and appropriately as a tool, just as they need to learn to read a clock, interpret a map, measure with a ruler, or use coins.

At this grade level, student work focuses on recognizing, interpreting, and using symbols on the calculator keys to do addition and subtraction. Some students may also use the calculator for multiplication and division. Implicit in their understanding is being able to assess a problem and determine how to use the calculator as a tool for solving it.

When should students use calculators? You will need to decide how calculators will be available to students. In our experience, when calculators are not readily available, or when there are only two or three in a classroom, they become desired objects and the choice to use a calculator is more focused on having control of the object rather than choosing it as a tool for solving a math problem. Also, when calculators are first introduced, there is an initial flurry of excitement. You may find that students choose to solve a problem using the calculator that they are capable of solving in their head. We have found that in general, students move to a point where they discover that there are many prob-

lems they can do in their head and see the power of the calculator as helping them work with larger numbers.

There will be times when students ask "What should I push?" as they try to solve a problem on the calculator. As with any material, being able to make sense of the problem and making decisions about how to solve it is the center of the experience. The response will probably be the same whether a student is using a calculator or cubes or pencil and paper. Knowing "what to push" is an important part of the solution, and once the buttons are pushed, looking critically at the answer and deciding whether it makes sense is the essence of the task.

The calculator does not substitute for work with materials or for the development of strong mental arithmetic strategies. Students must develop all of these strengths. You will need to make sure students get experiences with all of these separately and also have opportunities to choose from among any of these methods to solve problems.

We discourage thinking of the calculator as a tool to "check" with, as if other methods are somehow more fallible than the calculator. Any method can be used to check any other. In fact, it's as easy to make a mistake using a calculator as doing a problem mentally, with pencil and paper, or with materials. Throughout this curriculum, we encourage students to solve computation problems in more than one way in order to check their accuracy. If they have solved a problem using one mental arithmetic strategy, they might also solve it using another mental arithmetic strategy. If they have solved it with materials first, they might then solve it with a calculator. If they have solved it with a calculator, now they can try it mentally.

Continued on next page

Access and storage It is important for students to understand that calculators are an available math material, just like pattern blocks or interlocking cubes. They need to know where they are stored, how to take them out, and how to put them away. If you have a calculator for each student, and if you think your students can accept responsibility for the calculator, it is helpful for students to keep their calculators with their individual materials so that they get used to using them.

It may not be practical to have calculators kept by individual students. One teacher stored the calculators in boxes, numbered each calculator, and assigned a number to each student. This way, each calculator was used by one child. This system not only gave the students a sense of ownership but also helped the teacher keep track of the calculators. A variation of this system could be used if fewer calculators are available, by assigning one calculator to a pair or a small group of students.

Close to 20 and Beat the Calculator

What Happens

Students are introduced to two activities that involve adding two or more numbers. For two sessions they work on both of these activities during Choice Time. At the end of Session 9, the class discussion focuses on addition combinations that are related to combinations of doubles. These are referred to as doubles plus or minus 1. Their work focuses on:

- developing strategies for addition combinations
- looking for relationships among addition combinations

Start-Up

Today's Number Suggest that students use doubles in their number sentences. For example, if the number they are working on is 36, possible combinations include: 18 + 18, 9 + 9 + 9 + 9, 10 + 10 + 8 + 8. Add a card to the class counting strip and fill in another number on the blank 200 chart.

Materials

- Number cards, without wild cards (1 deck per pair)
- Student Sheet 10 (at least 2 per student)
- Student Sheet 11 (1 per student for homework plus 6–7 for the classroom)
- Beat the Calculator Cards (1 deck per pair)
- Interlocking cubes (about 20 per student)
- Calculators (1 per pair)
- Chart paper
- Clipboards (1 per student, optional)

Activity

Close to 20

Tell students that you are going to introduce two new activities, both of which involve number strings. The first one is the card game Close to 20. You will need a deck of number cards (as described under Materials).

Demonstrate this game by dealing five cards to yourself and five cards to "the class." Have a student read your cards and write those numbers on the board under your name. Have another student read the cards for "the class" and record those on the board.

Ms. Jones	The Class
5 9 4 8 6	7 9 7 6 5

The object of this game is to choose three of your cards that total as close to 20 as possible. Let's look at my cards first. I think I'll choose the 9 and the 8. 9 + 9 is 18 so 9 + 8 is 1 less, that's 17. I need 3 more to get to 20. I'll add the 4. That's 21.

Show students Student Sheet 10, Close to 20 Score Sheet, and demonstrate how to write the three cards you used and your total. (You may want to sketch part of the score sheet on the chalkboard so students can see how you are recording.) Then explain that your score for this round is 1 because you were 1 away from 20.

I'll write a 1 in the score column and I'll take one cube. Now let's look at your cards. Who has an idea about which numbers we could choose to get close to 20?

Ask for two or three ideas for combinations of numbers. Again, demonstrate how to record the three numbers used and the total amount on the score sheet. If students chose a combination of exactly 20, explain that score.

If you score exactly 20, then your score is zero and you do not take any cubes. In this game the person with the fewest cubes at the end of the game is the winner. When everyone has had a turn, each player discards the three cards that were used and gets three new cards.

Demonstrate how to discard the used cards and deal three more new cards to you and to "the class." Record the new numbers on the board and play one more round as a class.

Explain that this game will be available for students to play during Choice Time.

Activity

Beat the Calculator

Beat the Calculator is similar to the number strings students have been working on. This is a partner activity, and each pair will need a calculator, a deck of Beat the Calculator Cards, and paper for each student.

In this activity one partner is going to use the calculator to solve the number string. The other partner is going to solve the problem by looking for combinations of numbers and adding them in his or her head. Each person has to write down the total on a piece of paper. If you disagree, then you'll have to find a way to check what you have done.

Demonstrate the activity with a volunteer. Decide who will use the calculator first. Choose one problem from the deck and write it on the board. Ask someone to say "Go," and then you both solve the problem. Compare your answers, then each of you explain how he or she solved the problem. If you disagree, model how you might check the total.

Each person should record the problem he or she solved and the answer. Partners should take turns using the calculator and solving the problem mentally.

Choice Time

For the remainder of this math class and for most of tomorrow, students will be working on either Close to 20 or Beat the Calculator. They should work on each activity at some point in the next two class periods. Post the following choices:

> 1. Close to 20
>
> 2. Beat the Calculator

Choice 1: Close to 20

Materials: Decks of number cards, without wild cards; Student Sheet 11; Close to 20 (Directions); Student Sheet 10, Close to 20 Score Sheet; interlocking cubes

Two to three players are each dealt five cards. Players each choose three cards that total as close to 20 as possible, then record the cards and total on their score sheet. Their score for the round is the difference between the total and 20. They record their score and take that number of cubes. The three used cards are discarded and three new cards are dealt. The winner is the player with the fewest cubes (the lowest score) after five rounds.

Choice 2: Beat the Calculator

Materials: Calculator, decks of Beat the Calculator Cards

Students work in pairs to solve the number string problems. They take turns solving the problem mentally and using a calculator. They compare their totals and discuss their strategies. Each person records the problem and the total on paper.

Observing the Students

These Choice Time activities focus on using familiar or known number combinations to add groups of numbers. As you observe, try to get a sense of how students are adding. Do they look at the whole problem for familiar combinations, or do they add the numbers consecutively? If they add consecutively, do they use fingers or materials and count all the objects, or do they count on from one number? If students are using familiar combinations, which combinations are they comfortable with? Are they able to use these to figure out related combinations (double plus or minus 1 or combinations related to the 10's)?

Close to 20

- How do students select their three numbers? Do they select them randomly, or do they have a strategy such as choosing all the highest numbers or choosing two high numbers and then comparing that total to 20?
- Do they add two numbers, compare the total to 20, and then decide on its value?

Beat the Calculator

- How comfortable and accurate are students using the calculator?
- Are students able mentally to compute any portion or all of the number string? Are students using familiar combinations to do so?
- Can students explain how they solved the number string?
- How do students double-check when their totals do not agree?

At the end of each session, after students have cleaned up their materials, remind them to reord what they have done on their Weekly Log.

At this time, you may wish to collect the classroom sets of Number Cards and store them for use later, in Investigation 4.

Activity

Class Discussion: Combinations Related to Doubles

Near the end of Session 9, bring the class together for a discussion. Students will need paper and pencil and you will need to record either on chart paper or on the chalkboard. Students need a surface to write on, either at a table or desk, or a clipboard or book.

Begin the discussion by having students generate the double combinations and their totals from 1 + 1 to 10 + 10. List these on the board.

We have been talking about strategies for adding two or more numbers. Lots of times knowing how to solve one problem can help you solve another problem. For example, how could knowing 4 + 4 help you solve 5 + 4 ?

Record 4 + 4 and 5 + 4 on the chalkboard and ask for ideas. Then ask students how knowing 4 + 4 could help them solve 3 + 4. Record 3 + 4 on the chalkboard as shown:

$$3 + 4 = 7$$

$$4 + 4 = 8$$

$$5 + 4 = 9$$

All three of these problems are related. Can you see how?

Some students will use 4 + 4 and add 1 (doubles plus 1), and other students will use 4 + 4 and subtract 1. Some students may notice that the totals increase by 1. There will also be students in the classroom who do not yet see the relationship between these combinations. Choose another double and a related combination to list on the board.

Here's another set of problems that is similar. How does knowing 6 + 6 help you solve 5 + 6?

Once again, list the double and the two related problems on the board and ask students if they see how these three problems are related. List their observations on the board so that you can refer to them later.

$$5 + 6 = 11$$

$$6 + 6 = 12$$

$$7 + 6 = 13$$

Working in pairs, have students generate other combinations that are related to the doubles. They should list the combination and the related double on their paper.

After about 5 minutes, ask students for combinations and record them on the board.

Now we have examples of doubles and combinations related to doubles on the board. In the first part of our discussion someone noticed [*mention one of their observations*] that one of the totals is 1 more than the double and the other is 1 less than the double. Is that true for other doubles and their related combinations? [*Pause while students respond.*] Why do you suppose that is true?

Encourage students to explain their thinking about this statement. There will be a variety of explanations. Some students will see that in the three combinations one number remains constant while the other either increases or decreases by 1. Other students might see the increase or decrease in the total and then justify why that is so.

Some students will be able to visualize this by looking at the number sentences, whereas others may see it more clearly when it is modeled with concrete objects. You may want to use cubes to model a set of related combinations.

Sometimes we refer to these problems as doubles plus 1 or doubles minus 1. Why might that be a good name for these types of problems?

This is a convenient way to label these addition combinations. As problems come up throughout the year make reference to doubles plus or minus 1.

Note: You may want to record all the doubles and their related combinations on chart paper and post it in the classroom for students to refer to.

Sessions 8 and 9 Follow-Up

Homework

After Session 8, students play Close to 20 with someone at home. Each student will need Close to 20 Score Sheet, Close to 20 (Directions), and a deck of number cards. Since these are the same cards students used for Tens Go Fish and Turn Over 10, students should already have a deck at home.

Today's Number and the Magic Pot

What Happens

As a class, students generate number sentences for Today's Number. They try to use combinations of 10, doubles, or doubles plus 1 in their expressions. During the second part of the session, students solve a Magic Pot problem that can be used for assessment. Their work focuses on:

- expressing numbers in more than one way
- solving a problem involving doubling
- writing about problem solving

Start-Up

Today's Number Add a card to the class counting strip and fill in another number on the blank 200 chart. Students will generate combinations for Today's Number in the first activity of this session.

Playing Close to 20 at Home Students briefly discuss their experience of playing Close to 20 with someone at home.

Materials

- Chart paper
- Interlocking cubes (about 50 per pair)
- Student Sheet 14 (1 per student)

Activity

Post Today's Number on the board or chart paper. Ask volunteers to offer number sentences for this number and record these on the chalkboard. Then, focus students on generating expressions of Today's Number using combinations of 10, doubles, or doubles plus 1. As an example, list the following expression on the board. (In this example, 38 is Today's Number.)

Today's Number: Using Number Combinations We Know

$$10 + 7 + 9 + 3 + 9$$

Here's one way that I thought of to make Today's Number. Let's look at the whole problem. Who sees a combination of numbers you know, such as a way to make 10 or a double?

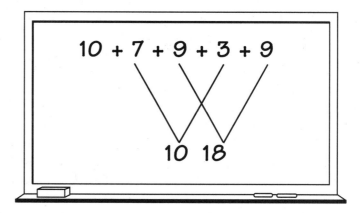

When students offer 7 + 3 as a combination of 10, connect the two numbers with lines and write 10. Do the same for 9 + 9. Many student will also identify 10 + 10 as a double.

We were able to make this problem simpler by looking for groups of numbers that go together. That's an important strategy when solving number problems. Think about another way of making 38. Can you use a combination of 10 or a double or doubles plus 1 in your number sentence?

Students should do this mentally. After a minute, accept one or two ideas from them, each time pausing to find familiar combinations. Then ask students to work on generating their own combinations for Today's Number. Each student should write the number on a sheet of paper and then list the combinations below. Have interlocking cubes available.

As students are working, circulate around the room to get a general sense of how easy or difficult this task is for them. It is important to note that for some, breaking down a number into smaller addends is more difficult than adding a string of numbers. The more often students work on activities such as these the more familiar they will become with the activity.

After about 10 minutes, have a few students share their number sentences. If many students are interested in sharing, use some time later in the day for this. Students should put these papers in their math folders.

Pose the following problem to the class:

Who remembers what happened in the story of the magic pot when both the man and the woman fell into the pot? What do you think would happen if our entire class fell into the magic pot? First write about how you would solve the problem. Then, write about what would be one good thing and one hard thing about having a double class.

Distribute Student Sheet 14, Our Class and the Magic Pot, to students. Explain that each student should solve the problem independently, then write an explanation about how he or she solved the problem. The written explanations should include words, numbers, or pictures. Then they should answer the questions, What would be good about having a double class? and What would be hard about having a double class?

❖ **Tip for the Linguistically Diverse Classroom** Suggest that students who have limited English proficiency draw pictures to show how they solved the problem. You might want to pair these students with those who are proficient in English to help them with their drawings.

You may need to adjust this problem for some students. If the number of students in the class is too large, suggest a smaller number. Those who complete the problem easily can pretend that this new "double class" fell back into the pot. How many students would there be now?

As students finish, ask them to read their responses to you. Without indicating whether their solution is correct, point out places where it is unclear. If they have not included words and numbers in their explanation, have them do so. As you look at their papers, consider the following:

■ How did students solve the problem?

■ How did they use materials? Did they count on from 1? from any number?

■ Did students break the numbers into more friendly numbers, such as 10's?

■ If students combined the two numbers, how did they do that?

Have the class share strategies for solving the problem. When all students who want to have shared, focus the discussion on one or two particular strategies that students used. For example, you might ask students to focus on the strategy of counting on, or pulling apart numbers into more manageable parts and then recombining them. See the **Teacher Note**, Assessment: Our Class and the Magic Pot (p. 64), for some examples of how students have solved this problem.

Assessment: Our Class and the Magic Pot

In Our Class and the Magic Pot, students imagine that their entire class fell into a magic pot that doubled the number of people in the class. Students' approaches to solving this problem can give you information about their strategies for doubling numbers and more specifically their strategies for combining two-digit numbers. The following examples of student work provide a range of responses that you might expect in a second grade classroom.

Lionel's Response:

Lionel drew a picture to solve the problem. He represented each person in the class with a stick figure, drawing two rows of 29 figures each. Lionel also represented the problem numerically as a problem involving two equal groups, as evidenced by the equation 29 + 29 that he wrote. In observing Lionel's work, his teacher noted, "Lionel counted his figures by 1's, beginning at 1."

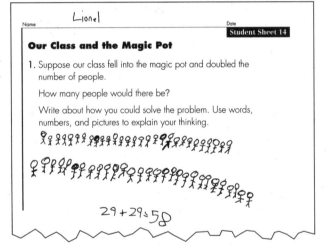

Many students used a strategy similar to Lionel's, although they used cubes to represent the number of people in the class rather than drawings. Many students built two towers of 29 cubes and then counted the cubes by 1's. A few students counted on from 29. In every case students understood the concept of doubling and could represent and solve the problem.

Ebony's and Paul's Responses:

Ebony and Paul used a class list to help them solve the problem. Instead of doubling the entire group as Lionel did, Ebony and Paul doubled each person, noting this by putting two tally marks next to each name on the list. They then counted these marks by 2's. Paul included his teacher in his list (using 30 people instead of 29) and found that there would be 60 people.

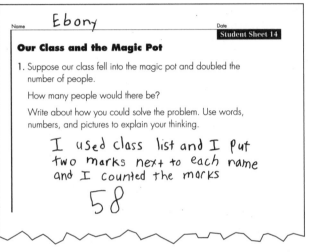

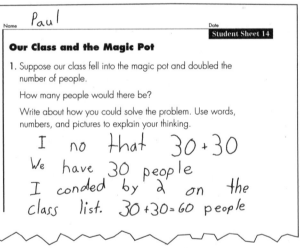

Because of the way Ebony and Paul chose to solve the problem, counting by groups of 2 naturally emerged as part of their strategy.

Continued on next page

Ebony and Paul are thinking of doubling in a slightly different way from Lionel. As you observe students in your classroom, you will probably find some who solve this problem by doubling the whole, while others double each individual part of the whole. You might want to compare strategies for this problem to strategies used in the next investigation, when students determine the total number of legs in their classroom.

Chen's Response:

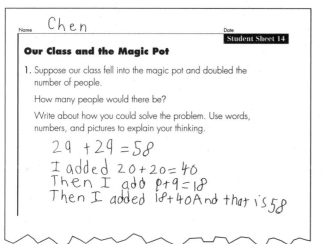

This student is breaking numbers into parts and then recombining the parts in order to get the total. Chen provides a detailed explanation. He breaks the 29 into a group of 20 and a group of 9 and then adds 20 + 20 to get 40 and 9 + 9 to get 18. He then combines these two parts to arrive at 58.

Ayaz's Response:

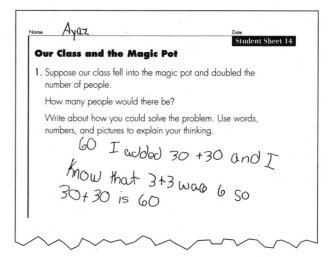

Ayaz also included his teacher in the count when he doubled the number of people in his class. Ayaz used something that he already knew, 3 + 3 = 6, and related this to the larger problem 30 + 30 = 60. Encouraging students to use what they already know to solve new problems is a powerful problem-solving strategy.

One purpose of assessment is to get a sense of how individual students are working with specific ideas and concepts. Another purpose can be to get a clearer sense of how the class as a whole is approaching these same ideas. After you have looked at each student's work, you might want to sort this set of papers by students who used a similar problem-solving strategy. In this way you can get an overall picture of your class. This information might influence how you focus discussions or highlight specific strategies or ideas in future sessions.

Counting Pockets

Materials

- Interlocking cubes (about 5 per student)
- Large jar
- Pocket Data Chart (new or from previous pocket activity)
- Class list of names
- Previous Choice Time materials (Close to 20, Beat the Calculator)

What Happens

Students collect data about the number of pockets worn by people in the class. As a group, they post the data on the board and look for familiar combinations of numbers, such as combinations of 10's, to help them add the number of pockets worn by the class. During the second half of the session, students complete Choice Time activities. Their work focuses on:

- collecting data
- using familiar number combinations to add a total

Start-Up

Today's Number Sometime during the school day, students brainstorm ways to express the number of days they have been in school. Suggest that students use both addition and subtraction in each expression. Also, add a card to the class counting line and fill in another number on the blank 200 chart.

Activity

Counting Pockets

The routine How Many Pockets? gives students an opportunity to collect, represent, and interpret numerical data through an experience that is meaningful to them. As students collect data about pockets throughout the year, they create natural opportunities to compare quantities and to see that data can change over time.

How Many Pockets? is one of three classroom routines that occur regularly throughout the *Investigations* curriculum. The complete write-up of this routine, which includes several versions, can be found at the end of this unit (p. 144). If you are doing the full-year grade 2 *Investigations* curriculum, students will be familiar with this routine and you should proceed with the following activity. If this is your first *Investigations* unit, familiarize yourself with this routine and do the basic pocket activity *instead* of the following activity.

Collecting Data: Post the data chart from previous Pocket Days where students can see it.

How many pockets are we wearing today?	Pockets	People
Pocket Day 1	74	26

Ask students to look at the data and decide if today the class is wearing the same, more, or fewer pockets than before. Encourage students to give their reasoning, which may include attendance or a change in the weather.

How many pockets do you think all the people in our class are wearing today?

Record students' estimates on the chalkboard or on chart paper. Pass around interlocking cubes and have each student take the same number of cubes as he or she has pockets. Then collect these cubes by having students place them in a large jar. Mark a line on the jar to indicate how full of cubes the jar is. This information will be used during a later pocket session.

Today we are going to collect pocket data. When I call your name, tell us how many pockets you have and I'll record that information on this large class list of names.

Beginning at the top of the list, ask students for their pocket information. Record the number of pockets next to their name. (Use large numerals so students can see them.) Collect data from everyone in the class. Before you continue, ask every student to check that the number next to his or her name represents the correct number of pockets.

Today we are going to count pockets by adding the numbers on this list. Take a look at our pocket data. Before we start adding, look at the whole problem. Think of this as one giant number string. OK, let's look for combinations of 10.

As students point out combinations of pockets that make groups of 10, draw lines on the chart connecting these numbers and write a 10 at the intersection. As numbers are used, cross them out so that students do not get confused. Some students might mention combinations of three or four numbers that make 10. Your chart may look like the diagram that follows.

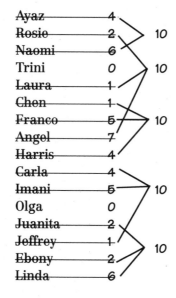

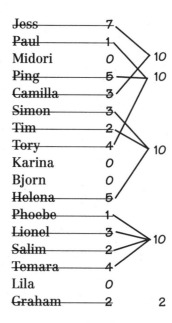

$$10 + 10 + 10 + 10 + 10$$

$$10 + 10 + 10 + 10 + 2$$

$$90 + 2$$

92 pockets

When no more combinations of 10 can be made, help students identify which numbers have not been added and again ask how these numbers could be combined. There may be discussion about how to count the zeros if you had some people without any pockets.

When all of the original pocket numbers have been added, focus students' attention on adding together the groups of 10 that they have made. Some students might want to count these by 10's. Finally, add any remaining numbers to calculate the total number of pockets.

One purpose of doing this activity is to provide students with one model for adding large groups of numbers. The technique of joining familiar combinations will make sense to some students. Don't insist, though, that every student use this type of organization when adding numbers.

Checking the Count   As a way of checking the number of pockets, count the cubes in the jar. Ask students to suggest strategies for counting the cubes. Choose one strategy and use it to count. If the two totals do not correspond, check both the count and the addition.

Enter today's pocket information on the class Pocket Data Chart and compare the data. Are students wearing more or fewer pockets as time goes on? Why might this be?

Note: Save the jar and the mark showing the height of the cubes collected for a later pocket session.

Finishing Choices

For the remainder of the session students finish working on activities from the previous Choice Time. Post the following choices:

> 1. Close to 20
>
> 2. Beat the Calculator

For a review of the descriptions of Choice 1: Close to 20, and Choice 2: Beat the Calculator, see pages 57–58.

Observing the Students

As you observe, try to get a sense of how students are progressing with their adding. Do they seem to be familiar with more combinations than before? Are they able to add without counting each unit? Are they using more strategies than before? Are students progressing with their accuracy in using the calculator? How comfortable do they seem with computing numbers mentally?

At the end of the session students should record what they did on their Weekly Logs and put all papers in their math folders.

INVESTIGATION 2

Grouping by 2's, 5's, and 10's

What Happens

Session 1: Using Groups of 2 Students count the number of legs in their classroom. They compare the number of people and the number of legs, look for patterns in the numbers, and try to describe a general rule for finding the number of legs for any group of people. For homework they write a riddle that describes the number of people and pets that live in their house.

Session 2: Exploring Multiples of 5 Students exchange and solve the People and Pet Riddles they wrote for homework. As a whole group they count around the class by 5's, use the class Hundred Number Wall Chart to keep track of their count, and then discuss the patterns they see.

Session 3: Counting by Different Groups Students work in pairs to express a number using only multiples of 5 and then explore counting a group of objects in more than one way. They are introduced to tallies as a way of keeping track of groups of 5.

Sessions 4 and 5: Counting Choices Students work on activities involving counting and grouping. In addition to Counting Bags, students are given a counting problem that involves calculating the total number of fingers in their classroom.

Session 6: Ways to Make 15¢ As a whole class students examine all the different ways to make 15¢. They share coupons they have brought from home, find different ways to sort them, and discuss the coin values of coupons.

Sessions 7, 8, and 9: Coins and Coupons Students work on four different activities, all involving coins and counting. They match coins to coupon values, find groups of coupons that total specific amounts, play a money game called

Collect 50¢, and explore all the possible ways to make 25¢ using pennies, nickels, and dimes.

Session 10: Collecting Pocket Data Students collect data about the number of pockets worn by the class. They are introduced to the Hundred Number Wall Chart as one way of counting the number of pockets. Students continue working on Choice Time activities.

Mathematical Emphasis

- Developing counting strategies
- Exploring patterns and developing fluency in skip counting by 2's, 5's, and 10's
- Exploring 5 and its multiples
- Becoming familiar with the relationship between skip counting and grouping (for example, as you count 5, 10, 15, you are adding a group of 5 to the total each time)
- Exploring ways of recording and keeping track when counting large groups
- Becoming familiar with coin equivalencies
- Using money as a model for counting by 5's and 10's

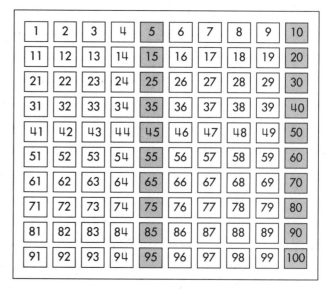

What to Plan Ahead of Time

Materials

- Interlocking cubes: class set (All sessions)
- 5"-by-8" cards or paper: at least 1 per student (Session 1)
- Overhead projector (Session 1, optional)
- Class list: 3 per student (Sessions 1, 4, and 5)
- Hundred Number Wall Chart, number cards and transparent pattern markers (Sessions 2, 3, and 10)
- Chart paper (Sessions 2, 3, and 6)
- Shoe box or container (Session 2)
- Counters such as buttons, beans (Session 3)
- Resealable plastic bags or envelopes: 40 (Sessions 3, 6, and 9)
- Containers for coin sets (Sessions 6–9)
- Coupons (Sessions 6–9)
- Plastic coin sets (real coins may be substituted), 30 pennies, 20 nickels, 20 dimes, 20 quarters: 1 per 3–4 students (Sessions 6–9)
- Number cubes with numbers or dots 1–6: 2 per 2–3 students (Sessions 7–9)
- Large jar (Session 10)
- Pocket Data Chart, from previous session (Session 10))

Other Preparation

For Session 1

- Prepare a class list of student names. Make 3 copies per student.
- Duplicate Student Sheet 15, 1 per student.
- Collect store coupons from newspapers or store flyers for use in Sessions 6–9.

For Sessions 2 and 3

- If you do not have the manufactured Hundred Number Wall Chart from the grade 2 *Investigations* materials kit, you can make one using heavy poster board, paper fasteners, and white and colored index cards. Attach 100 paper fasteners evenly across the poster board in 10 rows of 10. Cut 50 white index cards in half and punch a hole in the top of each one. Number the cards 1–100 and hang each card from a paper fastener. Cut the colored index cards in half and punch a hole in the top of each one. Number the cards from 1–100. These can be used to show a pattern in place of the transparent pattern markers.
- Prepare bags of 20–70 assorted small objects, such as pennies, beans, cubes. Letter each bag for identification purposes. Make 1 bag per pair plus a few extra.

For Sessions 4 and 5

- Duplicate Student Sheet 16, 1 per student.

For Sessions 6–9

- Gather store coupons. Organize the coupons in small plastic bags or envelopes, 10–15 per bag, 1 bag per pair.
- Duplicate Student Sheet 17, 1 per student.

For Session 10

- Use jar from previous pocket activity with mark to show height of cubes collected.
- Pocket Data Chart with data collected from previous sessions will be used here.

Using Groups of 2

Materials

- Interlocking cubes (about 50 per pair)
- Class list (at least 1 per student)
- 5"-by-8" cards or paper (1 per student for homework)
- Student Sheet 15 (1 per student for homework)
- Overhead projector (optional)

What Happens

Students count the number of legs in their classroom. They compare the number of people and the number of legs, look for patterns in the numbers, and try to describe a general rule for finding the number of legs for any group of people. For homework they write a riddle that describes the number of people and pets that live in their house. Their work focuses on:

- counting by groups of 2
- looking for patterns in multiples of 2

Start-Up

Today's Number   For each number sentence, check to see if any of the numbers can be expressed as doubles. For example, if the number is 42 and one number sentence is 40 + 2, each of these numbers can be written as doubles of other numbers: 20 + 20 + 1 + 1. The twenties can then be broken down into 10 + 10 + 10 + 10 + 1 + 1, etc. Add a card to the class counting strip and fill in another number on the blank 200 chart.

How Many Legs Are in Our Class?

Present this situation to students:

Suppose I wanted to figure out the number of legs in our classroom. How could I do that?

As students offer suggestions, list their ideas on the chalkboard. (If necessary, clarify that you want to find the number of people's legs, not chair legs or table legs.)

Ask students to suggest methods for solving the problem but not actually solve it yet. Students might suggest counting; using two cubes for each person then counting cubes; drawing pictures; or using the number of people in the class (29, for example) and doubling it (29 + 29), connecting this problem to the Magic Pot problem. By encouraging students to share their ideas you not only acknowledge that there are a variety of ways to think about this problem but you also may broaden students' ideas about how this problem can be solved.

Make sure that students have access to materials such as interlocking cubes, which they can use to solve the problem. If they are not accustomed to using a class list, introduce this as an available material.

There will be many times when we will be collecting information about the people in our class. I have made a list of all the people in our room. How might you use this list to solve the problem?

Note: If you have a classroom pet, students might include it in the count! Discuss this as a class and decide if and how to include this information.

When all ideas have been shared, tell students that they can solve this problem either alone or with a partner, but each student will need to write an explanation of his or her problem-solving strategy. The explanations should have a title and include words, numbers, and pictures. Put the following prompt on the board.

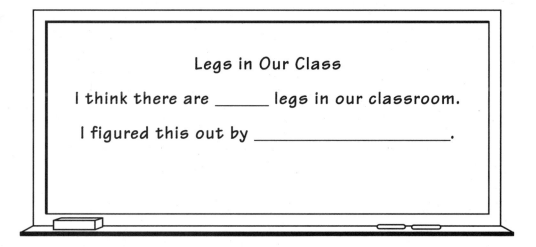

Legs in Our Class

I think there are _____ legs in our classroom.

I figured this out by _____.

Students can check their work by solving the problem another way.

After students have finished, ask for volunteers to share their solutions by reading their explanations and showing their work. If students used materials to solve the problem, suggest that they share these as well.

Counting Around the Class   As a way of double checking the number of legs in the classroom, have students count aloud. Beginning on one side of the classroom, count the legs of the people in the room by 1's. As students count, encourage them to say the odd number softly and the even number louder so they can hear the counting-by-2 sequence and see that it relates to counting a group of 2 objects.

Quite often students will know the first five or six numbers when counting by 2's but lose count when the numbers get into the teens. What is most important in this experience is that they to see that when counting by 2's (or any other group), they are really counting groups of 2 objects.

Looking for Patterns

Write the following on the chalkboard:

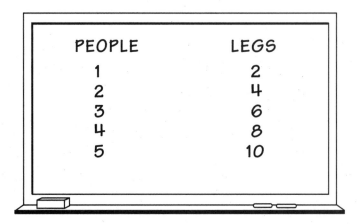

PEOPLE	LEGS
1	2
2	4
3	6
4	8
5	10

Many times you can find patterns in numbers. Take a look at this chart and talk with someone sitting near you about any patterns you notice, such as the number of people and the number of legs they have.

Encourage groups of students to talk together about what they notice in this chart. By having students share in pairs or small groups, they have an opportunity to share their ideas and listen to the ideas of others. After a few minutes call the groups back together and have some students share their observations. Record these next to the chart.

Suppose there were 9 people. How many legs would there be?

Some students will know right away that there would be 18 legs. Pose another problem, such as, **What if there were 26 legs? How many people would there be?** Encourage students to generate rules for solving these problems.

Students' strategies will vary considerably. Some students will count each leg, others might count by groups of 2, and some will double the number of people. What's most important is that students have a way of solving problems that makes sense to them and that they begin to explore number patterns and begin counting by groups of objects in a meaningful way. **The Dialogue Box,** What Do you Notice About These Numbers? (p. 76), will give you an idea of the range of strategies that exist among second graders.

Homework

For homework students will write a riddle about the number of people and pets that live in their home. Write the following riddle on the chalkboard.

> There are 12 legs in this group.
>
> There are 60 fingers in this group.
>
> There are 12 eyes in this group.

Here's a riddle about a group of people who live together. See if you can figure out how many people are in this group.

Discuss with students which clues were most helpful and how they know for certain that this is a riddle about six people.

Distribute Student Sheet 15, People and Pet Riddles, and a 5"-by-8" card or piece of paper to each student. This riddle is more complicated than the previous one because it includes a pet. Explain to students that for homework tonight they will solve the riddle on the student sheet and then make up a riddle about the people and pets that live in *their* home. Each of their riddles should include information about the number of legs, eyes, and fingers in their house. They should write their riddle on the 5"-by-8" card.

❖ **Tip for the Linguistically Diverse Classroom** Suggest that students with limited English proficiency use numbers and pictures to write their riddles. For example:

Give students the option of making up more than one riddle. After they have written a riddle about the people in their home, they could make up a riddle about any group of people and pets.

If there is time, students can begin this activity at the end of this session.

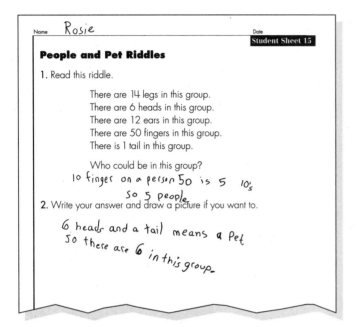

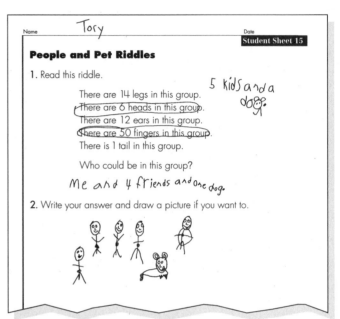

D I A L O G U E B O X

What Do You Notice About These Numbers?

In this discussion during the activity How Many Legs Are in Our Class? (p. 72), a variety of solutions and observations are shared. The discussion continues as students look for patterns on a chart listing the number of people versus the number of legs. Students are encouraged to make a general rule for determining the number of legs based on their observations.

How could we find the number of legs that there are in our classroom?

Karina: Count them!

How could we count them?

Chen: We could count by 2's because the number of legs that each person has is 2.

Naomi: You could also double the number of people in the room. Like 26 + 26, because one of the 26's would be for one of everybody's legs and the other 26 would be for the other leg.

Karina: You could figure out the number of legs in each group and then add up those numbers. But if you want to include the teacher you would add 2 more to the total.

Simon: You could go around the room and count everybody's legs by 1's.

Continued on next page

continued

How would we know if we counted everyone?

Simon: Once you count someone they could cross their legs and then you would know.

That's a good way of keeping track. Finding a way to keep track of what you have counted and what you still need to count is an important part of counting.

Salim: I would use cubes. Everyone could take two cubes and then we could put the cubes together and count the cubes.

If everyone made a tower of two cubes, what would each cube represent?

Salim: The person's legs.

And what would one tower represent?

Salim: The person?

So if we made towers, how many towers should we have?

Salim: 26. No, 27 if we include you!

Ayaz: You could also use a class list to keep track. I'd put a 2 next to each person's name and then add up all the 2's.

After the class has determined the number of legs in the classroom, they discuss the patterns that they see in the chart recorded on the board.

What do you notice about these numbers?

Samir: The numbers on the legs side are even, and on the people side they are even and odd.

Helena: They are in order from smallest to largest.

The number of people goes up by 1 and the number of legs goes up by 2. Why do you think that happens?

Naomi: When you add a person you are adding only one thing but you are adding two legs.

Karina: You count people by 1's and legs by 2's.

Rosie: You are doubling the amount every time. You put in a 1 and 2 comes out. You put in 3 and 6 comes out.

Think about some of the ideas that have just been shared. Is there a way to predict the number of legs for any number of people?

Rosie: Like I said, just double the number of people. Say if you had 25 people there would be 50 legs.

Karina: I'd just add people plus people.

Identifying patterns in numbers is a powerful mathematical tool, and using these patterns to make a generalization is an important aspect of mathematical thinking.

PEOPLE	LEGS
1	2
2	4
3	6
4	8
5	10

Exploring Multiples of 5

Materials

- People and Pet Riddles from last night's homework
- Hundred Number Wall Chart with number cards and transparent pattern markers
- Interlocking cubes (5 cubes per pair)
- Chart paper
- Shoe box or container

What Happens

Students exchange and solve the People and Pet Riddles they wrote for homework. As a whole group they count around the class by 5's, use the class Hundred Number Wall Chart to keep track of their count, and then discuss the patterns they see. Their work focuses on:

- solving riddles involving groups of 2's, 5's, and 10's
- counting by groups of 5
- looking for patterns on the 100 chart when counting by 5's
- combining multiples of 5 to equal a total

Start-Up

Today's Number   Sometime during the day, students generate expressions for Today's Number using only addition and only two addends. Encourage students to look for patterns in these expressions. For example, for the number 42: possibilities include 40 + 2, 39 + 3, and 38 + 4. Add a card to the class counting strip and fill in another number on the blank 200 chart.

Activity

Solving People and Pet Riddles

Have students exchange the People and Pet Riddles that they wrote for homework. Working in groups of three or four, students should exchange cards, solve the riddle, and explain to each other their problem-solving strategy. Encourage students to point out clues in the riddles that are confusing or unclear so that the author may correct them.

As students are solving the riddles, circulate around the class and listen as they explain their problem-solving techniques. Make note of riddles to share with the class, then have students read these aloud and ask others to tell how they would solve the problem.

Collect students' riddles and over the next few days write two or three riddles on the board for the class to solve in their free time. At the end of each day you can discuss the riddles and share solutions.

Counting Around the Class

For this activity everyone needs a partner. If there are 28 students in the class, how many pairs of students will we have?

Whenever you need to organize students into working groups, enlist their help to figure out how many groups or teams there will be. When students are paired up, distribute interlocking cubes so that each pair has a train of 5 cubes. As a group, students will count the total number of cubes.

Note: There are two reasons why this activity is done in pairs. One is for peer support and the other is to keep the number of cubes under 100. If you have a small class, have each student contribute a tower of 5 cubes.

Every pair of students in our class has built a train of 5 cubes. About how many cubes do you think we have altogether? [*Record estimates on the chalkboard.*]

We are going to collect your cube trains in this box. As cubes are added we will keep count of how many are in the box. I'll keep track of the count by placing markers in our Hundred Number Wall Chart.

Have one student pass around the box. Highlight the fact that with each pair of students, 5 more cubes are being added to the total amount. In this way you help students to see that counting by 5's is connected to accumulating groups of 5 objects.

OK, so Simon and Franco put 5 cubes in, then Olga and Chen added 5 more. That makes 10 cubes. How many will there be when Ebony and Karina add their tower of 5?

Take students' suggestions. For example:

Yes, 10 plus 5 is 15 in all. 5, 10, 15 cubes. Let's keep adding.

The class as a whole counts by 5's as trains are added to the box. Occasionally stop the count and ask students how many pairs have contributed trains. Check this by counting the number of trains. As students are counting, keep track of the count by placing transparent squares in the Hundred Number Wall Chart to highlight the counting-by-5's number sequence. Continue, asking students to predict what number will come next. When all cubes have been collected, discuss the chart.

We just counted the number of cubes in the box by groups of 5. What do you notice about this sequence of numbers?

Some students are likely to notice that all the counting-by-5 numbers end in either a 5 or a 0, that they form a pattern on the chart with the 5 and 10 columns highlighted, and that every row has two numbers marked off. These observations begin to give students a sense of 5 and its relationship to 10, and that patterns exist in counting sequences.

Students might be interested in exploring the following problem:

In the last problem *every pair* of students put a train of 5 cubes into the box and we ended up with [75] cubes. If this time, *every person* puts a train of 5 cubes into the box, how would the total number of cubes change?

Students may say that there should be more cubes in the box, and some may even be able to predict that the number of cubes will double. Ask students to give reasons for their predictions, then try the activity.

Note: Leave the transparent pattern markers in place on the chart for the next activity. (See Session 3.)

Counting by Different Groups

What Happens

Students work in pairs to express a number using only multiples of 5 and then explore counting a group of objects in more than one way. They are introduced to tallies as a way of keeping track of groups of 5. Their work focuses on:

- combining multiples of 5 to equal a total
- counting a set of objects in more than one way
- recording their counting strategies
- using tallies to count groups of 5

Start-Up

Today's Number   Add a card to the class counting strip and fill in another number on the blank 200 chart. Students will generate combinations for Today's Number in the first activity of this session.

Materials

- Hundred Number Wall Chart with multiples of 5 highlighted
- Interlocking cubes (about 50 per pair)
- Chart paper
- Prepared counting bags (1 per pair)

Activity

Students find different ways to express a number that is a multiple of 5 using only multiples of 5. For example, 10 + 5 + 5 = 20; 25 − 5 = 20. Plan to spend about 20 minutes on this activity and the remaining class time on Counting Bags.

Note: If you are keeping track of the number of school days and today's number is a multiple of 5, use this number. Use yesterday's or tomorrow's number if one of those is a multiple of 5.

As a class, generate a few examples and list these on the chalkboard or on chart paper. Call students' attention to the counting-by-5 numbers that have been highlighted on the Hundred Number Wall Chart. (Transparent pattern markers were placed on the chart in Session 2.)

Who can think of a way to make the number 50 using only the numbers that are highlighted on the chart? You can use addition or subtraction or both in your number sentence.

Today's Number

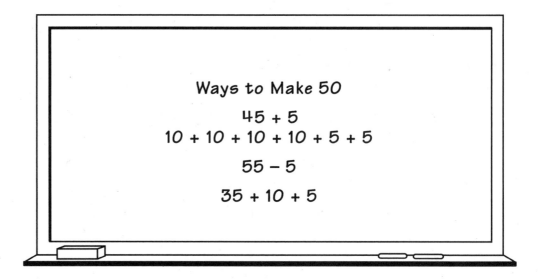

Ways to Make 50

45 + 5

10 + 10 + 10 + 10 + 5 + 5

55 − 5

35 + 10 + 5

List three or four expressions on the board to make sure students understand the task. Students then work with partners to list all the different ways they can to express Today's Number using only the multiples of 5 (or the counting-by 5-numbers).

As you circulate around the class, notice how students approach this task. Are they able to break apart numbers into groups of 5's or 10's? Do they use addition and subtraction? Do any students use multiplication or division? Do they begin with one equation and then break it down into smaller groups of 5's or 10's?

You might suggest that students use cubes to help them model combinations for the number. If some students are having difficulty, work with them as a group on this task while others work independently.

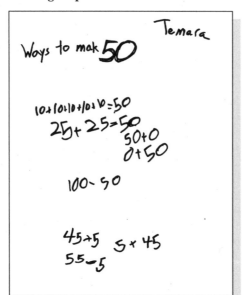

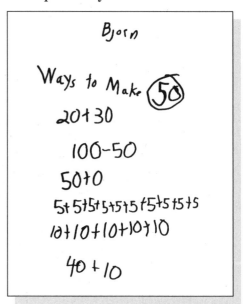

After about 10 minutes, gather students together and record some of their expressions on the chalkboard. Ask a few students to share their approach to generating expressions for Today's Number using only multiples of 5.

Breaking numbers apart into smaller components and then being able to put them back together is important in developing number sense. Because 5's and 10's (and their multiples) are important numbers in our number system, students should become familiar and comfortable with these groups. This same activity can be repeated whenever Today's Number is a multiple of 5. In this way students build on these initial experiences.

Activity

Counting Bags

Display a bag of objects and ask students to suggest how they might count them. Some students may suggest counting them one by one while others will probably suggest organizing them into equal groups such as 2's, 5's, or 10's.

Explain that each pair of students will have a bag of objects to count. Partners should agree on two different ways to count the objects. When they are finished counting, students write how many objects were in their bag and how they counted. Point out the letter on each bag and tell students to include this information also.

Give each pair a bag of materials and paper. Observe students as they work, noting the types of counting strategies they use and their explanation of how they counted. If some pairs finish early, suggest that they exchange bags and check each other's counts.

When most students have finished, ask volunteers to share how they counted the objects in their bags. As they share, record their counting strategies on the board using words and/or numbers. This models for students ways to record their verbal explanations. For example: "I counted by ones: 1, 2, 3, 4, 5, 6, all the way up to 36." "I made groups of 5 and then counted, 5, 10, 15, 20 . . . " "I put the cubes into towers of 2 and then counted 2, 4, 6, 8 up to 28."

Some students will use other methods of representation, such as drawings or tallies. To discuss tallying, point out students who counted by 5 or suggest this as another way of counting.

Some of you made groups of 5 to count the objects in your bag. Sometimes when people are counting by groups of 5 they use tally marks to keep track of their count.

Demonstrate how to make tally marks (four slashes with the fifth one a diagonal line) as you count a few objects. Emphasize that each slash stands for one object and that the diagonal line stands for the fifth object in the group.

Let's make tally marks to represent the number of people in our class. With your partner, see if you can figure out how many bundles of 5 to make to represent each class member.

Give students a few minutes to work on this problem. Notice how they keep track of the number of slashes they have made. Do they count by 5's and then add on the extras?

Call the class back together and have students share how many groups of 5 they made and how many extra slashes there were. If this is a difficult activity for students, have everyone stand up and count off by 1's. As they say their number they can sit down as you record the count on the board using tally marks. Occasionally stop to ask how many have been counted.

Explain to students that in the next few days they will be working on two different activities that involve counting sets of objects and writing about how they counted. They will use the counting bags that they worked with today as well as solving a problem about the people in the classroom.

Counting Choices

What Happens

Students work on activities involving counting and grouping. In addition to Counting Bags, students are given a counting problem that involves calculating the total number of fingers in their classroom. Their work focuses on:

- counting by 2's, 5's, and 10's
- recording counting strategies

Start-Up

Today's Number   Ask students to use combinations of 10 in their number sentences. For example, if the number they are working on is 54 and one number sentence is 10 + 10 + 10 + 10 + 10 + 4, ask students if there is another way of making 10 such as 6 + 4 + 6 + 4 + 6 + 4 + 6 + 4 + 6 + 4 + 4. Add a card to the class counting strip and fill in another number on the blank 200 chart.

Materials

- Prepared counting bags from Session 3 (1 per pair)
- Interlocking cubes
- Counters
- Class list of names (1 per student)
- Student Sheet 16 (1 per student)

Activity

Choice Time

For the next two sessions students work on two Choice Time activities: Counting Bags and Counting on Our Fingers.

You have been solving lots of problems about our class. A few days ago you found how many legs were in our classroom. Now, during Choice Time, I'd like you to try a problem called Counting on Our Fingers. You will need to find the number of fingers that are in our classroom. After you solve the problem, you will write your answer on Student Sheet 16, Counting on Our Fingers, and explain how you solved it. Use words and numbers in your explanation, and if you made a picture or a chart, include that too.

Post the following list in the classroom. Explain that students should complete both activities by the end of tomorrow, preferably counting at least two of the counting bags.

> 1. Counting Bags
>
> 2. Counting on Our Fingers

Choice 1: Counting Bags

Materials: Prepared counting bags from Session 3

Students choose a bag of objects. Then they count the objects in two different ways and record their counting methods. Their explanation should include the total number of objects in the bag and should show how they arrived at that total. See the **Teacher Note**, Writing and Recording (p. 89), or suggestions about how to help students record their mathematical thinking.

Choice 2: Counting on Our Fingers

Materials: Interlocking cubes, counters, calculators, class list

Students calculate the number of fingers in the classroom. They record their solution and write an explanation of how they solved the problem. Remind students to attach any class lists or pictures that they drew to help them solve this problem.

This may be a challenging problem for many second graders. Depending on the needs of individual, students, you may want to limit the problem by suggesting they count all the right-hand fingers of people in the class, or they could choose a group of eight people and count the fingers in that group.

Observing the Students

These activities focus on counting and grouping strategies. In general, try to get a sense of how students are counting and if they use groups in meaningful ways. Some students may organize objects into groups of 2 or 5 but then count them by 1's. This suggests that students have not yet developed an understanding of how one number can stand for many objects. By continuing to provide students with many opportunities to organize concrete objects into groups and to listen to how classmates are counting, students will come to count groups in ways that are meaningful to them.

Use the followng questions to help guide your observations as students are working:

Counting Bags

- Do students have a way of keeping track of the objects they have counted?
- Do they count one by one?
- Do they recognize that they will get the same total whether they count by 2's or by 5's?
- Do they get the same total when they count in two different ways? If not, do they count again?
- How do they record what they did?

Counting on Our Fingers

- How do students organize this problem? Do they use materials such as cubes, a class list, or the calculator?
- How do they keep track of the fingers they have counted? Do they count by groups of 5 or 10?
- Can they work with numbers over 100? Do they add the numbers? If so, how?

At the end of each session, after students have cleaned up their materials, remind them to record what they have done on their Weekly Logs.

Leave at least 15 minutes at the end of Session 5 to have the following discussion with students.

Teacher Checkpoint

How Many Fingers?

Class discussions provide students with a way of sharing their strategies and listening to the strategies of others. For the teacher, class discussions are opportunities not only to assess how individual students approached a particular problem but also to gain a sense of where your class is as a whole with respect to particular concepts and ideas.

Most often there will not be time for every student to share how he or she solved a problem. One strategy to consider is to ask questions such as, "Carla decided to count by 10's as a way of solving this problem. Did anyone else use a similar strategy?" After acknowledging students' strategies, the question "Did anyone else use a different strategy?" helps to move the discussion forward. The **Dialogue Box,** How Many Fingers? (p. 90), is an example of how one second grade teacher led this discussion.

The activity Counting on Our Fingers provides students with a natural situation for grouping by 5's or 10's. As students develop the concept of larger numbers and of many to one, and eventually of place value, one of the ideas that they will need to consider is what types of groups to use and count by. At some point in the discussion, explore with students why they chose to count in the ways they did.

I'm noticing that many of you decided to solve this problem by using groups of 5 or groups of 10. Why did you choose those groups? Did anyone group by 2's or another number? Why or why not?

As students share, pay particular attention to how they kept track of the total number of fingers. Did anyone make larger groups? You may want to record some systems on the board for all students to see.

At the end of this discussion you may want to have students count off by groups of 10. Go around the classroom and as students count, have them raise both of their hands to indicate the addition of 10 fingers to the count. Stop at various points and check how many fingers have been counted. (You may want to group 10 students as a way of keeping track.)

So far we have 80 fingers. How many people have we counted? How many more fingers do we need to make 100? How many people is that? Let's have those people stand together over here.

Not all students will be ready to use groups of 10 in their problem solving or keep track of such large numbers. These ideas develop over time with many experiences. Students will continue to work on these ideas throughout this unit and future units.

Writing and Recording

Just as students should be engaged in frequent mathematical conversation, so too should they be encouraged to explain their problem-solving strategies in writing and with pictures and diagrams. Writing about how they solved a problem is a challenging task but one that is worth the investment. As with any writing assignment, many students will need support and encouragement as they begin to find ways of communicating their ideas and thinking on paper, but even the very youngest students can be encouraged to represent their problem-solving strategies.

The range of students' abilities to write and record varies greatly in any second grade classroom. In part this is because second grade students are just beginning to feel comfortable in the areas of reading and writing, and in part because reflecting on one's own thinking is a challenging task. Initially, some students will record a few words and possibly some numbers that describe their strategy, while others might be able to write considerably more. Encouraging students to draw pictures as part of their explanation is often a way into the task. Explain that mathematicians often write and draw about their ideas as a way of explaining to others how they are thinking about a problem. The more often students are encouraged (and expected) to write and record their ideas, the more comfortable and fluent they will become.

Students benefit tremendously from discussing their ideas prior to writing about them. Sometimes this happens in pairs and other times in large-group discussions. Questions such as "How did you solve the problem?" or " Can you tell me what you did after you put the cubes into groups of 5?" may help extend students' thinking. During whole-class discussion it is important to model writing and recording strategies for students so that they can see how their mathematical strategies could be recorded using words, numbers, and/or pictures. For example, when Jeffrey reported to the class how he solved an Enough for the Class? problem, he said: "I put the cubes in groups of 5 and then I counted 5,

10, 15, 20, 25, but there was 1 left over so that made 26."

His teacher recorded Jeffrey's strategy on the board. As she recorded, she reiterated his strategy in words.

> 5, 10, 15, 20, 25
> 25 + 1 = 26 cubes

By recording students' explanations on the board, you begin to build up some possible models for them to use as they write about their thinking.

As with any type of writing, providing feedback to students is an important part of the process. As the audience for your students' work, you can point out those ideas that clearly convey student thinking and those that need more detail. Quite often by having students read their work aloud to you or to a peer, they themselves can identify ideas that are unclear and parts that are incomplete. The **Dialogue Box**, What Does It Mean to Be Finished? (p. 114), describes a teacher's interaction with two students about the writing they have done to explain how they solved a problem.

DIALOGUE BOX

How Many Fingers?

Second graders usually know how to make groups of 10 and count by 10's by rote, but to find totals they revert back to counting each object by 1's. Students who are beginning to understand that one number can stand for many objects will advance from counting by 1's to counting by groups, knowing that this will result in the same total number of objects.

This discussion, which takes place during Sessions 4 and 5 (p. 71), highlights counting strategies and ways of recording and keeping track of a count. Students share their strategies for finding how many fingers are in the classroom. Many students used groups of 10 to solve the problem, but not all used these groups in meaningful ways.

One activity you have been working on was to find the total number of fingers in our class. I noticed that there were many different strategies for solving this problem. As each person shares a strategy, I'd like you to listen carefully and decide whether your way was similar or different. If you used materials to solve the problem, you should show us what you did as you explain your strategy.

Ebony: I used a class list and I put a 10 next to everyone's name for 10 fingers. Then I added up groups of 10. I knew that ten 10's was 100 so I circled groups of 10. I added 100 +100 + 100 + 20 and got 320 fingers.

So you used groups of 10 to solve this problem. Your idea of putting circles around ten 10's seems like a good way of keeping track of what you counted. Did any one else use a strategy similar to Ebony's? [*Three students raise their hands.*] **OK, how about a different idea?**

Jess: Well, I used 10's too, but what I did was make towers of 10 cubes. I made a tower of 10 for each person and then I counted the towers.

I have two questions for you. How many towers did you have, and how did you count them?

Jess: I had 32 towers because I counted only the kids in the class. Sorry, I didn't include you! And then I counted by tens . . . 10, 20, 30, 40, 50, all the way to 320. [*Jess shows his 32 towers to the class.*]

Trini: I kind of did it like Jess but I just went down the class list and said, 10, 20, 30, 40, until I ran out of names. And I got 310.

Simon: I sort of did it like Jess. I used cubes too, but I counted them by 1's. I ended up with 282, but I'm not sure that's right.

Why do you think that?

Simon: Well, because everyone got 310 or 320 and also I kept loosing track of the counting so I might have skipped a number or something.

Juanita: I counted by 1's too. It took a long time.

How many of you used a strategy similar to Jess's or Juanita's? [*Many students raise their hands. Some comment that they made groups of 10 but counted by 1's.*] **So we've heard a few different strategies. Are there any others?**

Olga: Well, I thought that since everyone had 5 fingers on each hand I would count by 5's. First I tried to look at the people in the class and count, but I kept forgetting who I counted. So I used a class list and I made two marks next to each person's name. That was for each of their hands. Then I just counted by 5's. I said, 5, 10, 15, 20, 25, 30, 35, 40 for every mark. And I got 260.

Olga, I know that you got a little stuck when you reached 100. Can you explain what you did?

Olga: Well, I got stuck and you helped me. So I drew a line under the name where I had 100, that was at Harris. Then I started over again and counted to 100 again and drew a line under that name. Then I just counted what was left. [*She shows her class list to the group.*]

Continued on next page

continued

So you counted by groups of 5 and then you grouped those 5's into groups of 100's. That's similar to what Ebony did when she made smaller groups into bigger groups as a way of keeping track. I'm noticing that most of you either counted by 1's or by groups of 5 or 10. How come you chose those numbers?

Lionel [*laughing*]: Because for most people that's what fingers come in. You have 1 finger, or 5 fingers on one hand, or 10 fingers in all.

Trini: It wouldn't make sense to count by like 4's or 6's, especially if you made towers of 10.

Jess: But you could count by other numbers except it would be hard to keep track.

Juanita: And I even counted by 1's because that's easiest for me, but it was hard to keep track.

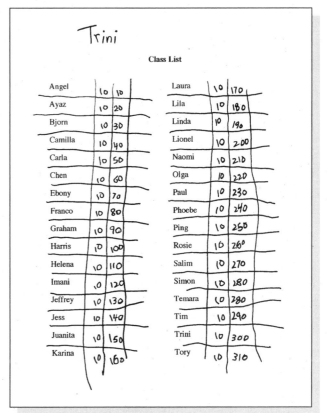

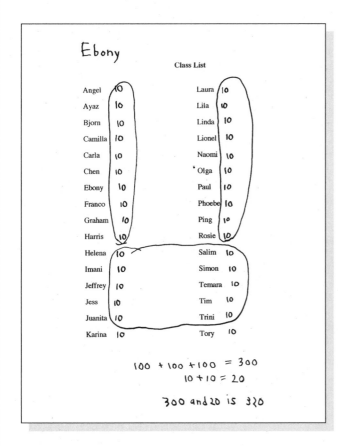

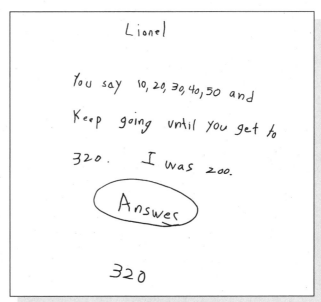

Ways to Make 15¢

What Happens

As a whole class, students examine all the different ways to make 15¢. They share coupons they have brought from home, find different ways to sort them, and discuss the coin values of coupons. Their work focuses on:

- becoming familiar with coin values
- finding all possible combinations of coins to equal 15¢
- matching coin combinations to cents notation

Materials

- Prepared bags with coupons (1 set per pair)
- Plastic coin sets in tubs or containers (1 set per 3–4 students)
- Chart paper

Start-Up

Today's Number   Suggest that students use doubles in their number sentences. For example, if the number they are working on is 36, some possible combinations are: 18 + 18, 9 + 9 + 9 + 9, or 10 + 10 + 8 + 8. Add a card to the class counting strip and fill in another number on the blank 200 chart.

Activity

Ways to Make 15¢

Distribute pennies, nickels, and dimes to small groups of students.

Suppose I went to the store and I bought something that costs 15¢. Decide with your group all the different ways I could make 15¢ using pennies, nickels, and dimes, and make those combinations on your desk. After you make each combination, write or draw it on paper. When you think you have all the possible combinations of 15¢, talk with your group about how you know for sure that there are no other possible combinations.

Allow students 10 to 15 minutes to work on this problem and then call the class together to share results. Record the combinations students suggest.

WAYS TO MAKE 15¢		
pennies	nickels	dimes
15	0	0
10	1	0
5	2	0
5	0	1
0	3	0
0	1	1

As students share, ask them to show how they recorded their combinations. Some students might draw each individual coin, others might use abbreviations such as 10 P + 1 N, and others might use coin notation such as 10¢ + 5¢. What is important is that they understand their system of recording and that their combinations do equal 15¢.

When students think that all combinations have been listed, ask them to explain why they are sure.

Tell students that the following day they will be working on an activity called Ways to Make 25¢. The activity is like this one except that they will be finding all of the ways to make 25¢ using pennies, nickels, and dimes.

This is a good time to introduce students to the quarter, as they will be working in combinations that equal 25¢ during the next three sessions. For now, distribute three to four quarters to each group of students and have them add these coins to their collection.

Matching Coins and Coupons

Distribute a bag of coupons to each pair of students.

I have put about fifteen coupons in each bag. Can anyone explain what a coupon is and how it is used?

Solicit a few ideas and then ask students to look at the amount of each coupon. List the amounts that they tell you on the board.

What do you notice about these numbers?

Some students will probably notice that they are all multiples of 5. Next, generate a list of the types of things that students have coupons for. List the items on the board.

Sometimes when people collect coupons, they like to organize then into certain types of groups. Take a look at our list. How might we organize these coupons?

List or group items that might belong to the same category or group, such as cleaning supplies, breakfast foods, things to drink, and so on.

Working in pairs, students look through all of their coupons and sort them into groups.

When they have finished, explain to students that they should choose one group of coupons and use coins to show the amount on each coupon. (Note: You might need to let two pairs of students share each coin set.)

Suppose I have a cereal coupon that is worth 35¢. How can I make 35¢ using pennies, nickels, dimes, or quarters?

Once students have tried this activity, tell them that this, along with Ways to Make 25¢, will be two of their activities during Choice Time.

Session 6 Follow-Up

 **Homework**

Students write five ways to show tomorrow's number using multiples of 5 and 10. If tomorrow's number is 50, students might write: 10 + 10 + 10 + 10 + 10 = 50 and 55 − 5 = 50. If tomorrow's number is not a multiple of 5, students can add on the extra number. Example: 51 = 10 + 10 + 10 + 10 + 10 + 1 or 51 = 50 + 1. If they wish, they can include a sixth way they got from a family member.

Coins and Coupons

What Happens

Students work on four different activities all involving coins and counting. They match coins to coupon values, find groups of coupons that total specific amounts, play a money game called Collect 50¢, and explore all the possible ways to make 25¢ using pennies, nickels, and dimes. Their work focuses on:

- becoming familiar with coin values
- finding all possible combinations of coins to equal 25¢
- matching coin combinations to cents notation
- finding combinations of coupons to equal a given amount

Start-Up

Today's Number   Collect or post the homework students did showing five or six ways to write Today's Number using multiples of 5 or 10. Add a card to the class counting strip and fill in another number on the blank 200 chart.

Materials

- Prepared bags with coupons (1 set per pair)
- Additional coupons
- Plastic coin sets in tubs or containers (1 set per 3–4 students)
- Number cubes with numbers or dots 1–6 (2 per 2–3 students)
- Student Sheet 17 (1 per student)

Collect 50¢ and Shop and Save

If you have completed the first unit in the grade 2 *Investigations* sequence, *Mathematical Thinking at Grade 2*, students will be familiar with the coin game Collect 25¢. Collect 50¢ is a variation of that game and is played with two number cubes. If students have not played Collect 25¢, briefly introduce the game to them. Based on their familiarity with money, Collect 25¢ may be an appropriate game for many students to begin with. In this game, players roll a number cube, collecting the number rolled in cents. Players can trade coins for equivalent amounts. The first player to collect 25¢ wins. It may be helpful to play a demonstration game with a volunteer.

The game Shop and Save is similar to Matching Coins and Coupons.

In this activity you need to find combinations of coupons that total a given amount. You will need a collection of coupons and Student Sheet 17, Shop and Save for this activity. One Shop and Save problem asks you to find three coupons for things you might need to buy if you were having a party. Write down which three coupons you would use and then find how much money you could save if you used all three coupons.

Another problem asks you to find two coupons that total exactly 50¢.

Distribute a copy of Student Sheet 17, Shop and Save, to each student and read through each problem. Because students will need to look for specific coupons, they should not be limited to the coupons in the bags but instead be able to choose from a larger collection of coupons stored in a box or tray. Students can record their work on the back of the student sheet.

❖ **Tip for the Linguistically Diverse Classroom** As you read aloud each problem on Student Sheet 17, Shop and Save, make sure students with limited English proficiency understand the idea of using coupons at the store. You might want to have students re-enact a situation at a market of making a purchase and using a cents-off coupon.

To be sure students understand the concept of doubling the coupon value, choose one coupon and ask students to show its value with coins. Then ask them to double the coin set (make a second matching set). Students can count all the coins to find the double value.

Activity

Choice Time

List the following activities and let students know that they will be working on these for the next three math sessions. Remind students that they do not need to complete every choice today. As with other Choice Time sessions, limit the number of students that can work on an activity at any one time. If there are activities you would like every student to complete, let them know at the beginning of Choice Time.

> 1. Ways to Make 25¢
>
> 2. Matching Coupons and Coins
>
> 3. Collect 50¢
>
> 4. Shop and Save

Choice 1: Ways to Make 25¢

Materials: Plastic coin sets (1 set per 3–4 students)

Students find all the possible ways of making 25¢ using pennies, nickels, dimes, and quarters. They make the combinations with coins and then record these on paper.

Choice 2: Matching Coupons and Coins

Materials: Prepared envelopes with coupons (1 envelope per pair), plastic coin sets

Students match the value of a coupon with a combination of coins that represents that amount. There is no recording sheet for this activity. Encourage partners to check each other's work.

Choice 3: Collect 50¢ (or Collect 25¢)

Materials: Plastic coin sets (1 set per 2–3 students), 2 number cubes

Players take turns rolling the number cubes and collecting the number rolled in cents. Players can trade coins for equivalent amounts. The first player to collect 50¢ wins. If students have not played this game before, have them begin by using one number cube and playing Collect 25¢.

Choice 4: Shop and Save

Materials: Prepared envelopes with coupons (1 envelope per pair); Student Sheet 17, Shop and Save (1 per student)

Students find combinations of coupons to fit each situation. They record the types of coupons used and the total amount saved on their Student Sheet. Depending on your students, you might suggest that each student do at least four of the problems on Student Sheet 17, making the remaining optional choices.

Observing the Students

As you oberve students working during Choice Time sessions, use the following questions to guide your observations.

Ways to Make 25¢

- Can students make 25¢ using combinations of different coins?
- Do they have a systematic way of keeping track of the combinations they make?
- How do they know when/if they have found all the possible combinations?
- How do they record their combinations?

Matching Coupons and Coins

- Can students identify the coin value of each coupon?
- Are they able to represent it using coins?

Collect 50¢

- Are students able to count out their coins accurately?
- Do they take their coins in pennies and wait until they collect 50¢, or do they trade in as they play?
- Do they trade in some coins for coins of equal value (two nickels for one dime)?
- Are students able to tell you how much their coins are worth?
- Do they know how much more they need to get to 50¢?

Shop and Save

- Are students able to find the right combination of coupons totaling the designated amount of money?
- Can students add coupon totals mentally, or do they count out coins for each coupon and then total the coins?

Five or 10 minutes before the end of each session, remind students to record what they worked on in their Weekly Logs. Suggest that they use the list of Choice Time activities that you posted as one reference for writing down what they did.

Activity

Class Discussion: Ways to Make 25¢

At some point during these Choice Time activities, discuss with students the possible combinations of coins they might use to equal 25¢. Encourage students to share strategies about how they solved this problem. Did they use a system for finding combinations, or was it more trial and error? Are they certain that they found all the possible combinations of coins? Why?

You might consider recording the combinations on a piece of chart paper and leaving the chart in the room for a while to see if students come up with any other combinations that total 25¢.

Collecting Pocket Data

What Happens

Students collect data about the number of pockets worn by the class. They are introduced to the Hundred Number Wall Chart as one way of counting the number of pockets. Students continue working on Choice Time activities. Their work focuses on:

- collecting and comparing data
- using the 100 chart as a counting tool
- using groups of 10 to count

Start-Up

Today's Number   Sometime during the school day, students brainstorm ways to express the number of days they have been in school. They add a card to the class counting line and also fill in another number on the blank 200 chart.

Materials

- Hundred Number Wall Chart (with number cards in place)
- Interlocking cubes (in a container)
- Large jar
- Pocket Data Chart (recorded in previous sessions)

Activity

Counting Pockets

Gather students together in a circle on the floor or large space in the classroom. Display the Hundred Number Wall Chart. Post the Pocket Data Chart showing the previous pocket counts. (This activity was first introduced in this unit in Investigation 1, Session 11, p. 66.) Review the pocket data that have been collected so far. Then ask students to estimate how many pockets the class is wearing today.

On our previous Pocket Day, we counted [*name number*] pockets. How many pockets do you think all the people in our class are wearing today?

Record students' estimates and ask them to share how they made their estimate. Some students may take the previous pocket data experiences into account. Others might glance at their classmates, getting some visual information about the number of pockets: "Most kids only have about 2 pockets and there are 25 kids in our class so I think about 50 pockets." Still others may simply guess.

Pass around a container of interlocking cubes and have each person take the same number of cubes as he or she has pockets. Call out numbers of pockets, asking students who have that many to put their cubes in the jar.

When all the cubes have been collected, ask students to compare how full the jar is today to the mark on the jar from the previous Pocket Day.

This mark shows how full the jar was on our last Pocket Day, when we had [*name number*] pockets. Look at the jar now. What can you tell about the number of pockets we have today?

Students may share descriptions such as "It's a lot more" or "We have just a few more [or fewer] pockets today."

Look at our Hundred Number Wall Chart and tell me what you notice about it.

Students may notice that each space on the chart has a number and that the numbers are in order from 1 to 100. Some might also notice and describe patterns (all the numbers with a 5 go down in a row).

We are going to use this chart to help us figure out today's pocket count. Let's fill the first row. A pocket can have only one cube, so we'll start by putting one cube in number 1's pocket, the next cube in number 2's pocket, and so on. How many cubes will we put in the first row?

One group of 10 cubes fits in this row. Think about your estimate. How many rows of the chart do you think would be filled by your estimate?

Note: If cubes do not fit in the pocket of the 100 chart, place the chart on the floor in the center of the circle and put the cubes on top of the numbers.

Invite students to share their predictions and their reasoning. This can provide you with information about whether any students use counting by 10's or can relate groups of 10 to a number. At this point in the year, many students may simply guess. For them, this activity provides an informal introduction to grouping by 10's as a way to count.

Have students count with you and help you place the cubes one by one on the chart. After you have put about half the cubes in the chart, examine it together.

How does the chart show how many pockets we are wearing today? How many rows did we fill? How could we use the rows to help us count the cubes?

Some students will most likely suggest counting by 10's. However, some may not and will need more experiences with counting and grouping. For each row, snap together the cubes to make a train of 10. As you do so, use the rows to encourage students to consider combining groups of 10. Record number sentences on the chalkboard as you add on groups of 10.

10 and 10 more is 20.   10 + 10 = 20

20 plus 10 is 30.   20 + 10 = 30

Together, count the trains by 10, counting on the remaining loose cubes.
Record the day's pocket total on the Pocket Data Chart. Compare today's
pocket data with previous data.

**Suppose we had used the Hundred Number Wall Chart to count our last
pocket data. Discuss these questions with someone sitting near you: How
many rows would we have filled on the Hundred Number Wall Chart?
How do you know?**

Ask a number of students to share their thinking. Encourage them to be
specific about how they thought about the problem. By having students verbalize their strategies you gain some insights into how they are thinking
about numbers in the 10's.

Using 10's develops over time through experiences with number, counting,
and grouping. This is only one of the many experiences with counting and
grouping that students will have in second grade. As the pocket routine
continues throughout the school year, watch for students' increasing ease,
ability, and understanding in counting and grouping by 10's.

For the remainder of this session students finish working on the Choice
Time activities presented in this investigation. You may feel that students
would benefit from continued work with some of the activities. The games
and activities could be available during free time, at indoor recess, or even
adapted for homework. As with most experiences, students benefit from
repeated opportunities to interact with materials.

**Finishing
Choices**

Introducing Addition and Subtraction Situations

What Happens

Session 1: Introducing Combining Situations
Students are introduced to combining problems. They solve problems and record their solution so that someone else can understand it. Several strategies are shared and recorded.

Session 2: Combining Notation In pairs, students share the combining problem they wrote for homework. As a class they make up problem situations for the expression 21 + 14. For most of the session they work on solving combining situations and writing a problem represented by combining notation.

Session 3: Introducing Separating Situations
Students are introduced to separating problems. Similar to the previous sessions, they discuss a story problem, solve a few separating problems, and record their solution so that someone else can understand it. Several strategies are shared and recorded.

Sessions 4 and 5: Making Sense of Addition and Subtraction Students solve a variety of story problems involving combining and separating, using their own strategies. Their job is to solve each problem, check their solutions, and clearly record their approaches. As a whole class, students discuss the relationship between addition and subtraction situations.

Mathematical Emphasis

- Developing models of addition and subtraction situations
- Solving problems using numerical reasoning
- Recording solution strategies clearly
- Considering the relationship between addition and subtraction
- Working with notation for addition and subtraction
- Matching addition and subtraction notation to situations they could represent

What to Plan Ahead of Time

Materials

- Overhead projector (Sessions 1 and 3, optional)
- Counters, such as interlocking cubes or tiles: about 30 per student (Sessions 1–5)
- Envelopes, such as legal size or large brown: 7 (Sessions 4 and 5)
- Paste or glue sticks (Sessions 4 and 5)

Other Preparation

For Session 1

- Think about situations familiar to students that you might use as contexts for addition and subtraction problems. Throughout this unit, you may want to substitute problems of your own. For more information, see the **Teacher Note**, Creating Your Own Story Problems (p. 108). If you use your own problem contexts, create the following kinds of problems: a combining situation (with one or two follow-up problems) for Sessions 1 and 2, a separating situation (with one or two follow-up problems) for Session 3, and a mixture of six to eight combining and separating problems for Sessions 4 and 5.
- Duplicate Student Sheet 18, Story Problems, Set A (or copies of the problems you have created), 1 per student.

For Session 2

- Duplicate Student Sheet 19, Story Problems, Set B (or copies of the problems you have created), 1 per student.

For Session 3

- Duplicate Student Sheet 20, Story Problems, Set C (or copies of the problems you have created), 1 per student.

For Sessions 4 and 5

- Duplicate Student Sheet 21, Story Problems, Set D (or copies of the problems you have created), 1 per student.
- Prepare story problems on Student Sheet 21 by cutting apart the sheet into individual problems. Store each problem in a separate envelope marked with its number. Paste an example of each problem on the front of the envelope so students can see which problem they are choosing.

Introducing Combining Situations

Materials

- Counters
- Student Sheet 18 (1 per student)
- Overhead projector (optional)

What Happens

Students are introduced to combining problems. They solve a problem and record their solution so that someone else can understand it. Several strategies are shared and recorded. Their work focuses on:

- combining quantities
- recording strategies

Start-Up

Today's Number Suggest that students use combinations of 10 in their number sentences. For example, if the number they are working on is 54 and one number sentence is $10 + 10 + 10 + 10 + 10 + 4$, ask students if there is another way of making 10, such as $6 + 4 + 6 + 4 + 6 + 4 + 6 + 4 + 6 + 4 + 4$. Add a card to the class counting strip and fill in another number on the blank 200 chart.

Activity

Problems About Combining

As students are introduced to story problems, try to refrain from labeling them ahead of time as "addition" or "subtraction." See the **Teacher Note,** The Relationship Between Addition and Subtraction (p. 118), for more information on ways that students interpret problems. For these five sessions, students will be presented with addition problems that involve finding a total amount by combining two quantities and subtraction problems that involve starting with a whole quantity and then removing or separating a part of that quantity. These ideas are expanded upon in a later unit, *Putting Together and Taking Apart,* in which students will also be introduced to a variety of subtraction situations including comparing two quantities.

Write a problem such as the one that follows on the chalkboard or overhead. Keeping the same numbers and the basic structure of the problem, you might want to create a problem that has a more familiar or timely context for your students. See the **Teacher Note,** Creating Your Own Story Problems (p. 108).

Here's a problem that I would like everyone to think about for a minute. Don't solve it yet. Just think about what is happening in the problem.

There are 12 children playing tag on the playground. Then 10 more children join the game. How many children are playing tag now?

Close your eyes and imagine what is happening. OK, who can explain what you see?

It is important that students be able to visualize what is happening in a problem before they attempt to solve it. After a couple of volunteers have shared their thinking, pose the following:

Will the answer to this problem be more than 12 or less than 12? Explain your thinking.

By encouraging students to think about the problem in this way, you allow them to think about what happens when two quantities are combined. Although the answer may seem obvious to those who understand combining situations, it may not be clear to all second grade students.

Solve this problem in whatever way makes sense to you. Use interlocking cubes or anything else you need. I would also like you to keep track of how you solved it on paper. Write or draw to show how you solved the problem so that someone else can understand it. You can use words, pictures, and numbers. Someone should be able to look at your recording and understand just what you did to solve the problem. When everyone has finished, we'll share strategies.

If you are using the problem above, give Student Sheet 18, Story Problems, Set A, to each student. If not, provide students with copies of the story problem that you have created. You may want to put a container of counters in reach of all students.

❖ **Tip for the Linguistically Diverse Classroom** As a comprehension aid for students with limited English proficiency, read the story problems aloud, asking students to draw a picture over key words as they listen to the problem.

Students work individually on the problem, but they should be seated in groups so that they can discuss their strategies. Remind them that their job is to solve the problem and to record it clearly—using words, pictures, and/or numbers—so that someone else can understand what they did. During this investigation, work with students to help them understand what it means to record their strategies.

Encourage students to solve this problem in *whatever way makes sense to them.* Some students may count all: They count out two piles of objects, put them together, then count the whole pile, starting with 1. Others may count on: They begin with one of the quantities and continue counting. Some students may approach this problem with numerical reasoning strategies, separating 10's and 1's or counting on by 10's. For more information about how students develop addition and subtraction strategies, see the **Teacher Notes,** Students' Addition and Subtraction Strategies (p. 109) and Developing Numerical Strategies (p. 110).

As you watch students, encourage them to describe their strategies clearly. You may need to ask them to add to or revise what they have written or drawn. Students who solve this problem easily and record their strategy well should find a way to check their answer. Ask students who are counting all or counting on to find a way to check their counting so that they are sure they counted correctly. It is very easy for anyone to miscount; students need to find ways to be sure they are counting accurately.

If the numbers you are using seem too large for some students, modify the problem to use smaller numbers. If some students finish easily, have them try the same problem with larger numbers.

When students are finished, they can describe their solution to partners. Those who have time can work on problem 2. They record their strategy below the problem.

Notice the variety of approaches students use so that you can encourage those with different strategies to share their work with the whole class in the next part of the session.

Activity

Sharing Strategies

Gather the group together to share strategies. Have something to count with available for students to demonstrate counting-all or counting-on strategies. As students share strategies, record them on the chalkboard. The ways you record will give students models for recording their own work. Record each strategy using numbers so students can refer to it easily ("Mine is like the first way," "Mine is almost the same as the second way, but . . ."). It's probably not wise to label them with students' names because other students are likely to have approached the problem in similar ways, and all students will enjoy feeling ownership of a strategy.

Here are the ways one teacher recorded three different strategies:

1. xxxxxxxxxxx oooooooooo
Counted them all by 1's and got 22. Counted by 2's to check.

2. Started at 12 and counted on. Kept track of the 10 on fingers.

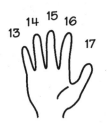 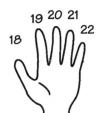

3. Thought of 12 as 10 + 2. 10 + 10 = 20; 20 + 2 = 22.

After several students have shared their strategies, ask:

Did anyone have a way that is different from the ways I've written here?

Before the discussion ends, ask each student to look at his or her own strategy and decide which of the ones you've recorded is closest to it. Ask students to raise hands to say which strategy they used. This is a way of validating all students' work and also giving you a sense of what kinds of strategies are being used in your class as a whole.

Session 1 Follow-Up

Students write and solve a story problem that's about combining two things, using something that they can see out of a window at home. Again, they need to show their solution strategy with some combination of numbers, words, and pictures.

🏠 Homework

❖ **Tip for the Linguistically Diverse Classroom** Suggest that students with a limited English proficiency use sketches and numbers to record their story problems.

Throughout this investigation and the next, students will be meeting addition and subtraction situations. You may want to use the sample story problems for this unit or you may want to create your own problems. Since you know which contexts reflect the interests of your students, and since you may need to adjust the numbers in problems for different students in your class, the following provides information about developing appropriate problems.

Creating Interesting Contexts   In creating addition and subtraction situations, use contexts that are familiar to students without being distracting. For example, the first problem in the investigation (see Student Sheet 18) is about a group of students playing tag. Change the problem to be about a playground situation familiar to your students.

Simple, familiar situations may be the most satisfying. One source of good situations is experiences that all students have had. For example, one urban class walks to a nearby park every day for recess; a problem based on that experience might be:

When we were at the park, I counted 26 squirrels on the ground and 12 more in the trees. How many squirrels were in the park?

Another class is having a series of bake sales at lunchtime to raise money for a class trip:

Jeffrey made 26 chocolate cupcakes and 12 vanilla cupcakes. How many cupcakes did we have to sell?

In one classroom, the teacher made up two characters, Ted and Sophia, who had experiences very much like those of her students, and built problem situations around these characters:

Ted and Sophia got a bag of peanuts. On the way home, Ted ate 26 peanuts and Sophia ate 12 peanuts. How many peanuts did they eat?

Many teachers also take advantage of special events, classroom happenings, seasons, or holidays for problem contexts:

If Rosie made 26 snowballs and Jess made 12, how many snowballs did they have?

Follow-Up Questions   Simple, familiar situations often suggest other problems that easily follow from the initial problem. These follow-up questions can be written down as options or they can be posed for students who have finished the first problem. Additional questions for the previous examples might include:

When I walked over, 10 of the squirrels ran away. How many were left?

We sold 22 cupcakes. How many were left over?

The next Friday we sold twice as many cupcakes. How many cupcakes did we sell?

Ted and Sophia started out with 50 peanuts in their bag. If they ate 38 peanuts, how many were left when they got home?

In making up follow-up questions, consider whether to keep the level of challenge the same or to make the problems easier or more difficult.

Adjusting Numbers in the Problem   You may have students who are comfortable working with numbers in the 20's, 30's, 40's, and higher, and other students who need to work with smaller numbers. Students who are primarily using counting strategies may lose track if the numbers with which they're working are too high.

Use the numbers suggested as a baseline problem and then adjust the numbers after students have solved the basic problem. In this way both you and students have some immediate information about the level of difficulty of the problem. Sometimes *you* can decide which students should use which numbers. At other times, let students choose appropriate numbers for themselves from some you suggest. Ask students to choose numbers that make the problem a little difficult for them. Let them know that different numbers will be challenging for different students and that it's important for everyone to work with problems that are right for him or her.

Students' Addition and Subtraction Strategies

As students develop strategies for problems about combining and separating, they are involved in two major tasks:

■ understanding the structure of the problem

■ developing numerical strategies for solving the problem

Understanding the Problem Structure To understand the structure of the problem, students need to form a mental image or build a concrete model of what is happening. For example, if the problem situation involves putting together Anna's 25 marbles and Jose's 11 marbles, students need to be able to imagine the two separate groups of marbles and the larger combined group. They need to develop a mental model of the actions and relationships.

Combining is a natural activity for young students. Visualizing problems that involve separating or taking away can be more difficult. For example, suppose the problem states that there are 25 birds on a fence and 11 flew away. The question is to find out how many birds are still on the fence. Some students begin to solve this problem by counting out or drawing *both* quantities, the 25 and the 11, then become unsure of what to do with them. In combining two quantities, the same action is performed on each quantity: Both quantities become part of a total accumulation. In removing or separating one quantity from another, however, one of the two quantities is a *part* of the other. The importance of visualizing the action is particularly critical as students model problems that involve taking one part out of a whole.

In order to visualize and describe the problem situation, some students will need to create a physical model of the actions. For example, one student might solve the problem about birds flying away by drawing 25 birds (or tallies), crossing out 11 of them, and counting the number not crossed out. Another student may count 25 cubes, remove 11 of them, and count the number remaining. These students are treating all the quantities in the problem as a collection of 1's

and using counting as their primary strategy. Eventually we want to move these students toward developing mental representations of the problem that allow them to develop strategies that involve working with groups.

Students who feel more confident visualizing the problem mentally will be able to use strategies that involve counting on or counting back. For example, a student might solve the marbles problem as follows: "Anna had 25 marbles, so I counted 11 more: 26, 27, 28, 29, . . . 35, 36." This student feels confident enough about visualizing the actions in the problem that she is able to hold the 25 in her head and add 11 to it, rather than counting out each quantity separately. Counting on—or counting back for subtraction—is a double-count situation. The student must simultaneously keep track of the numbers she is counting (26, 27, 28, . . . 36) and how many numbers she has counted (1, 2, 3, . . . 11). This double-counting becomes even more complex when the student is counting backward. If a student counts back to solve the bird problem, he must count down from 25 (24, 23, 22, . . . 14) while at the same time counting up to keep track of how many numbers he has counted (1, 2, 3, . . . 11).

Developing Numerical Strategies While students gain more experience in visualizing and modeling problem situations, these experiences help them learn about the structure of numbers. As they learn more about number relationships, they learn how to take numbers apart into useful chunks, manipulate these chunks, and then put them back together. For example, they see that 25 is composed of 20 and 5, or two 10's and 5, or 20, 3, and 2. They use this increased flexibility in thinking about numbers to solve problems using strategies that don't depend on counting by ones. See the **Teacher Note**, Developing Numerical Strategies (p. 110), for more specific descriptions of numerical strategies.

As students move from relying on counting by 1's to numerical strategies, researchers have found that students naturally develop strategies based on two ideas:

■ Numbers can be taken apart into pieces that are more convenient to work with.

■ It is often easiest to work with larger parts of the numbers first, then with smaller parts.

Students in the primary grades typically begin to develop strategies for combining and separating problems like these:

Anna had 25 marbles and Jose had 11 marbles. How many marbles did they have together?

1. $20 + 10 = 30$ **2.** $25 + 10 = 35$ **3.** $25 + 5 = 30$

     $5 + 1 = 6$ $35 + 1 = 36$ $30 + 6 = 36$

     $30 + 6 = 36$

In the first example, the student breaks up both numbers into 10's and 1's, adds the 10's first, then the 1's, then combines the two subtotals.

In the second example, the student starts with the largest number, adds on the 10's from the second number, then adds on the 1's from the second number.

In the third example, the student first breaks the 11 into 5 and 6 because she sees that she needs a 5 to get from 25 to 30. She adds on the 5 to the 25, then adds on the remaining 6.

There were 37 birds on the fence. 13 flew away. How many birds are on the fence?

4. $30 - 10 = 20$ **5.** $37 - 10 = 27$ **6.** $37 - 7 = 30$

     $7 - 3 = 4$ $27 - 3 = 24$ $30 - 3 = 27$

     $20 + 4 = 24$ $27 - 3 = 24$

In the fourth example, the student subtracts the 10's from the 10's, then the 1's from the 1's, then adds the results.

In the fifth example, the student starts with the first number, subtracts the 10's of the second

number, then subtracts the 1's of the second number.

In the sixth example, the student first breaks up 13 into 7 + 3 + 3 because he sees that it will be convenient to subtract 7 from 37 first. Then he subtracts each part of the 13, the 7, then the 3, then the other 3.

When adding, no matter how you break up the numbers, you always add everything back together to get the result. For subtraction, however, when parts are subtracted from parts, as in the fourth example above, the results are *added* together. Keeping track of what quantity you start with and what quantity is being subtracted, even when these numbers are broken up into parts, is a critical part of understanding subtraction. Here is one second grader's strategy for 37−13. This gives us some idea of the complex thinking involved in keeping track of the roles of each number in a subtraction situation:

> "Put the 7 from the 37 aside and save it. 30 minus 10 is 20. Then take away the 3 [from the 13] from the 20, so that's 17. Now add back the 7 you saved, and that makes 24."

These are all strong, mathematically sound strategies based on students' good understanding of the numbers and their relationships. They are using knowledge about how a two-digit number is composed of 10's and 1's, about what happens when you add or subtract a 10 from a two-digit number, and about combinations that make 10. These are the kinds of numerical strategies we want to encourage.

Students who learn to use strategies like these fluently and flexibly will eventually be able to solve any addition or subtraction problem. They will learn to modify these procedures to handle more difficult problems. For example, in the subtraction problem, *25 birds on the fence, 19 flew away*, the strategies must take into account that there are more 1's in the quantity being subtracted than in the original quantity (this issue is discussed further in the unit *Putting Together*

Continued on next page

and Taking Apart. Students will not need to learn the historically taught algorithms that use carrying and borrowing because they will develop their own algorithms based on sound understanding of the structure of numbers and operations.

About Notation As students are learning to record, some may write solutions using a string of equal signs, like this: 25 + 5 = 30 + 6 = 36.

This student is using the equal sign to indicate the sequence of operations being performed, but the equal sign is not used correctly here, since 25 + 5 is not equivalent to 30 + 6. Students may use symbols incorrectly as they work at developing their own strategies. When this occurs, model the correct usage for them.

Combining Notation

What Happens

In pairs, students share the combining problem they wrote for homework. As a class they make up problem situations for the expression 21 + 14. For most of the session they work on solving combining situations and writing a problem represented by combining notation. Their work focuses on:

- combining quantities
- recording strategies
- interpreting standard notation

Materials

- Counters
- Student Sheet 19 (1 per student)

Start-Up

Today's Number   Sometime during the school day, students brainstorm ways to express the number of days they have been in school. They add a card to the class counting strip and also fill in another number on the blank 200 chart.

Activity

Looking at Addition Notation

During the first 10 minutes of this session, group students in pairs and have them share the problems they wrote last night for homework. Each student should read the problem and then explain to a partner how he or she solved it. As students share, circulate around the classroom to get an idea of the types of situations they are writing about, the level of difficulty, and how clearly students were able to record their strategies.

Gather the group back together and call their attention to the addition problem you have written on the board: 21 + 14.

Think of a story that we could use for this addition problem. Imagine it in your mind. What is happening in the story that you imagine for 21 + 14?

Have three or four students share their stories. Explain to students that today they will write their story for this problem and then solve it. In addition, they will have two other combining problems to work on. Distribute Student Sheet 19, Story Problems Set B to each student. Read though the two problems to make sure students understand them. Explain that they should write a story for the given problem in number 3. Have paper available for students to record their work.

❖ **Tip for the Linguistically Diverse Classroom** It may help students with a limited English proficiency to continue to draw pictures over key words as they listen to you read the problems.

For the remainder of this session, students work on solving these three problems. Some students may want to work with partners, but remind them that each person needs to write an explanation of how he or she solved the problems. You may want to work with a small group of students who are having difficulty during this session.

Materials such as cubes and counters should be available to students. Encourage them to use materials to represent problem situations whenever possible.

In addition to becoming familiar with combining situations and developing strategies for solving combining problems, the emphasis of this investigation is for students to record their strategies in ways that are clear and understandable to others. Although some of the problems may seem easy for certain students, make sure that they are also able to record their ideas. Sometimes this is easier to do when numbers and problems are simpler. Emphasize that *being finished* means not only solving the problem but also writing a clear and complete explanation of how the problem was solved. See the **Dialogue Box,** What Does It Mean to Be Finished? (p. 114), for examples of a teacher helping students revise and expand their first attempts at recording their work.

Remind students to put all of their work in their math folders at the end of the session and fill out their Weekly Logs. If they have not completed every problem, explain that they will have more time to work on them during Sessions 4 and 5.

What Does It Mean to Be Finished?

In this classroom, the teacher is trying to help students understand what it means to record their strategies so that someone else can understand clearly how they approached the problem. This is very difficult for many students. Because their method seems so clear to them, they think that what they wrote or drew explains their thinking even when there are gaps in their recordings. Taking the point of view of a reader of their work is a new skill for students at this age and one that they will need to work on.

In this dialogue that takes place during the activity on p. 112, the teacher is giving students feedback on the following problem:

Luis and Kathy were collecting rocks. Luis found 16 rocks and Kathy found 24. How many rocks did the children collect? (Student Sheet 19, Problem 1.)

Carla [*reading her explanation to the teacher*]: 40 rocks. I used cubes and I counted.

OK, so that tells me how many rocks they collected and it tells me that you used cubes. How did you use the cubes to help you solve the problem?

Carla: Well, I counted out 16 and then I counted 24 more and altogether that was 40.

Do you remember how you counted the cubes to get 40?

Carla: I counted the tower of 16 from 24. Like I went 25, 26, 27, 28, and I ended up with 40.

So you counted on by 1's from 24?

Carla: Yes.

That sounds like a good strategy. See if you can go back and add those ideas to your explanation. Think about how you just explained to me what you did and write it down. Jeffrey, you're next.

Jeffrey [*reading his work*]: They have 40 rocks. I used my brain.

That's a good start, Jeffrey. Can you explain more about how you used your brain?

Jeffrey: I just thought 20 + 10 is 30 and 4 + 6 is 10. So 30 and 10 is 40.

I've heard you use that strategy before. Remember when we were working on a similar problem yesterday, we wrote down lots of people's strategies. Sometimes I used words or numbers to explain what people did. And sometimes I used both. How could your record your thinking so someone reading your paper would know just what you did?

Jeffrey: I'll try numbers.

Carla [*reads her revised work to the teacher*]: 40 rocks. I used cubes and I counted. I made a tower of 16 cubes and a tower of 24 cubes. Then I counted from 24. 25, 26, 27, 28, 29, 30, 31, 32, 33, 34, 35, 36, 37, 38, 39, 40. [*Carla has also added a picture of two towers, one with 16 cubes and one with 24.*]

That tells me much more information about what you did. I can understand how you solved the problem from this information. Before you write about the next problem, think about what you did. You can even explain it to a friend or to yourself to help you start writing.

When the teacher checks in with Jeffrey he has added the following to his recording:

$24 + 16 = 40$

$20 + 10 = 30$ $4 + 6 = 10$ $30 + 10 = 40$

i used my brane

Only by continuing to insist that students record their work fully can the teacher eventually help students imagine their audience.

Introducing Separating Situations

What Happens

Students are introduced to separating problems. Similar to the previous sessions, they discuss a story problem, solve a few separating problems, and record their solution so that someone else can understand it. Several strategies are shared and recorded. Their work focuses on:

- separating quantities into two parts
- recording strategies

Start-Up

Today's Number   Sometime during the school day, students brainstorm ways to express the number of days they have been in school using three addends. Add a card to the class counting strip and fill in another number on the blank 200 chart.

Materials

- Counters
- Student Sheet 20 (1 per student)
- Overhead projector (optional)

Introduce a problem such as the one below to the whole class. As in Sessions 1 and 2, you may want to change the context of the problem so that it is more familiar to students. The **Teacher Note,** Creating Your Own Story Problems (p. 108), provides suggestions for developing your own problems. Write the following problem on the board or overhead, then read it together with students.

Activity

Problems About Separating

> Andy had a bunch of 28 balloons. By mistake he let go of some and 15 of them flew away. How many balloons did Andy have left?

I would like everyone to think about this story problem for a minute. I don't want you to solve it yet. Close your eyes and imagine what is happening. OK, who can explain what you see?

Just as you involved students in visualizing combining situations, it is important for them to imagine what is happening in separating situations as well. This visualizing technique becomes very important when the types of situations become more complicated. For example: Andy had 28 balloons. By mistake he let go and some flew away. Andy had 13 balloons left . How many flew away?

After a couple of volunteers have shared what they think is happening, pose the following question:

Do you think that the answer to this will be more than 28 or fewer than 28? Why?

By encouraging students to think about the problem in this way, you allow them to think about what happens when you start with one amount and then some part is taken away or separated. Although the answer may seem obvious to those who understand separating situations, it may not be clear to all second graders. Subtraction situations are more difficult for students to follow and represent, making it extremely important for them to have opportunities to visualize problem situations and represent those situations with materials or pictures.

Give students Student Sheet 20, Story Problems, Set C.

I'd like you to solve the first problem in whatever way makes sense to you. You can use cubes or any other material. Record your work below each problem. When everyone has finished, we'll share strategies.

Remind students that their job is to solve the problem, to check the solution, and to record what they did in a way that someone else can understand. As you watch students work, remind them to record clearly. See the **Dialogue Box,** What Does It Mean to Be Finished? (p. 114).

Students should continue to have easy access to counting materials. They can work individually or in pairs, but each student records individually. Students who have time can do the second problem.

Meet together as a whole class to share solution strategies for this problem. Use this time as an opportunity to model ways of recording a variety of solutions. For example, here are some solutions that students are likely to offer.

1. □□□□□□□□□□□□□□□☑☑☑☑☑☑☑☑☑☑☑☑☑
   Counted out 28 cubes. Took away 15. Counted what was left.

2. Counted backward: 27, 26, 25, 24, 23, 22, 21, 20, 19, 18, 17, 16, 15, 14, 13

3. 28 − 10 = 18   18 − 5 = 13

4. Counted up: 16, 17, 18, . . . 26, 27, 28. Kept track of the number counted on fingers, like this:

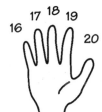
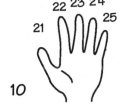
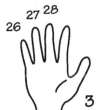

16 17 18 19 20   10

21 22 23 24 25

26 27 28 3

Some students may use addition to solve what you think of as a subtraction problem. See the **Teacher Note**, The Relationship Between Addition and Subtraction (p. 118). Some strategies that involve counting backward (Example 2) can be very confusing for students in that they "land" on 13 but fail to realize that 13 is included in the group they are separating, thus they are really left with 13 objects. This is an excellent example of when modeling a problem with cubes can help students see their mistake more clearly than trying to explain it in words.

Session 3 Follow-Up

Homework

Students write and solve a subtraction story problem: Start with some amount, then part of that amount is lost, taken away, eaten, or whatever. Ask students to base their problem on something they can see out a window at home. Students should be getting used to the requirement that they show their solution strategy with a combination of numbers, words, and pictures.

The Relationship Between Addition and Subtraction

It is easy for us to consider problems with similar solution strategies the same, yet they may actually appear quite different to the students. We may assume that certain situations are addition and others subtraction because we are used to thinking of them that way, but we may find that students solve problems in unexpected ways.

For this reason, as you introduce addition and subtraction problems to students, don't label them ahead of time as "addition" or "subtraction." A critical skill in solving problems is deciding what operation is needed. Further, many problems can be solved in a variety of ways, and students need to choose operations that make sense to them for each situation. For example, students may solve problems that you think of as subtraction by using addition (in fact, many adult problem solvers also do this). Consider the following problem:

Yesterday at the park, I counted 39 pigeons. When a big dog walked by, 17 of them flew away. How many were still there?

Most of us learned to interpret this situation as subtraction, and we may naturally assume that students should also see this problem as subtraction. Students who use counters to solve this problem will probably count out 39, remove 17, and count how many remain. However, there are many other ways to solve this problem:

- counting down from 39 (38, 37, 36, 35, . . . 24, 23, 22) and keeping track of how many numbers are counted

- counting up (18, 19, 20, . . . 37, 38, 39) and keeping track of how many numbers are counted

- counting up by 10's: 27, 37, that's 20, and 2 more is 39; or, from 17 to 37 is 20 and 2 more is 39

- counting down by 10's: 29, 19, that's 20, and 2 more down to 17 is 22

- using important numbers, such as multiples of 10: from 17 to 20 is 3, from 20 to 30 is 10, from 30 to 39 is 9; 10 + 3 is 13 and 9 more is 22 (or 13 and 7 more is 20 and 2 more is 22)

- subtracting 10's and 1's: 30 − 10 is 20; 9 − 7 is 2; 20 and 2 is 22

Some of these methods are based on subtraction (moving from 39 down to 17), but others are based on addition (moving up from 17 to 39). Which method is chosen may have to do with a person's mental model of the situation: Do you see this problem as a taking away situation to be solved by subtraction, as an adding on situation, or as a gap between two numbers that might be solved by either addition or subtraction, depending on which is easier in the particular situation? The important point here is that any of these methods are appropriate for solving this problem. Addition is just as appropriate for solving this problem as subtraction and, for many students, will make more sense.

Making Sense of Addition and Subtraction

What Happens

Students solve a variety of story problems involving combining and separating, using their own strategies. Their job is to solve each problem, check their solutions, and clearly record their approaches. As a whole class, students discuss the relationship between addition and subtraction situations. Their work focuses on:

- modeling problem situations
- selecting appropriate strategies for different problem structures
- recording strategies
- interpreting standard notation for addition and subtraction
- considering the relationship between addition and subtraction

Start-Up

Today's Number   Sometime during the school day, students brainstorm ways to express the number of days they have been in school. Suggest that they try using both addition and subtraction in each expression. Add a card to the class counting line and fill in another number on the blank 200 chart.

Materials

- Counters
- Paste or glue sticks
- Prepared envelopes containing Student Sheet 21 (1 problem per envelope)

Activity

During the first 10 minutes of this session, group students in pairs and have them share the problems they wrote last night for homework. Each student should read the problem and explain to a partner how he or she solved it. As students share, circulate around the classroom to get an idea of the types of situations they are writing about, the level of difficulty, and how clearly students were able to record their strategies.

Gather the group back together and call their attention to the subtraction problem you have written on the board: 30 – 13.

Think of a story that would be represented by this subtraction problem. Imagine both the story and the problem in your mind. What is happening in the story that you see to go with 30 – 13?

Looking at Subtraction Notation

Have three or four students share their stories. Students will have a chance to write their story for this problem and then solve it during today's class. In addition, they will have some other separating and combining problems to work on.

Show students the envelopes containing the story problems from Student Sheet 21, or the ones you have written. Explain that each story problem is stored in a separate envelope and that they can choose one problem at a time to work on. Students will need paper and glue sticks.

After you have chosen the problem you want to work on, paste it on paper. Use the space below each problem to show how you solved it.

Demonstrate for the class how to paste problems onto the paper. Depending on how much space individual students need, one or two problems can be pasted on one side of a page. Encourage students to work on one problem at a time so they can focus their attention on that problem and not feel as though they need to rush to finish many problems on one page.

You may want to distribute problem 7 on Student Sheet 21 (30 − 13) to students and have them begin by writing a story for that problem and solving it.

Over the two sessions, students will work on three to six problems. This is enough if they are to solve, check, and record carefully. You will need to keep track of which problems they work on to make sure each student works on both combining and separating problems.

Activity

Story Problems

For the remainder of this session and most of the next, students work independently on a variety of story problems representing different combining and separating situations. They can work individually or in pairs. Remind students that they should do three things:

■ Solve the problem.

■ Check their solution by using a different strategy.

■ Record their solution so that someone else can understand it. After recording their solution, they should check with at least one other student to see if someone else can understand their drawing and writing.

As students encounter new problem types, you may want to pull together small groups to help get them started. Encourage them to think through the whole problem before they try to solve it. They can use pictures or counters to model the problem for themselves. Discourage approaches such as relying on individual words in the problem that might seem to signal a particular operation. See **Teacher Note**, "Key Words": A Misleading Strategy, (p. 123).

Teacher Checkpoint

Student Strategies

By observing and/or looking at student papers, classify the strategies students are using. Look at one combining problem and one separating problem for each student. In particular, notice which students are counting by 1's to solve all problems, which students are building on number relationships they know, and which students are breaking up numbers in flexible ways. For example, you might review students' work on Student Sheet 21, problem 4.

The second grade class went on a trip to the zoo. There were 32 students and 12 adults on the trip. How many people went on the zoo trip?

Look for:

- Students who count out a group of 32 things and a group of 12 things, then count all from 1
- Students who count out a group of 32 things and a group of 12 things, then count on from 32
- Students who reason, using important number relationships; for example: 32 is close to 30 and 12 is close to 10. 30 + 10 is 40, then add on the 2 from 32 and the 2 from 12. 40 + 2 + 2 = 44
- Students who use the base-ten structure of the numbers; for example: 32 + 10 = 42; 42 + 2 = 44

Class Discussion: Is It Adding or Subtracting?

Sometime during Session 5, bring the class together to discuss their work. Focus the discussion on a separating problem such as:

There were 29 first and second graders on the bus. When 13 children got off the bus, how many children were left on the bus?

Ask students to think about this problem and share the strategies they would use to solve it. Record some of their suggestions on the chalkboard or chart paper. Then pose a related combining problem, using the same numbers for the two quantities:

There were 13 first graders and 16 second graders on the bus. How many students were there in all on the bus?

Ask students to talk in pairs for a few minutes about this problem. It is very likely that someone will observe that the two problems are related. If not, ask students whether they see anything about one problem that helps them solve the other. Continue the discussion, using some of the following questions:

■ How are these two problems related? What's the same about them?

■ Does one help you solve the other?

■ How would you write down the first one using numbers?

■ How would you write down the second one?

Some students may propose the same way to write both problems. The first problem might be recorded as $29 - 13 = 16$ or $13 + 16 = 29$. Some students might think of this problem as subtracting 13 from 29, while others might think of it as adding on to 13 to make 29. Ask students to reflect on what they mean by adding and subtracting:

Karina, when you say you added to do the first one, what do you mean? What does it mean to add? What does it mean to subtract? Do they have anything to do with each other?

The ideas presented in this session and this investigation may be quite new to students. In the following investigation, students will continue to solve and discuss strategies for solving combining and separating problems.

"Key Words": A Misleading Strategy

Some mathematics programs have advocated a "key words" technique to help students solve story problems. Students are taught to recognize words in a problem that provide clues about how to choose which operation to use to solve the problem. For example, *altogether* or *more* signal addition, whereas *left* or *fewer* signal subtraction:

I have 5 marbles. Sue gave me 6 more. How many do I have altogether?

I have 16 marbles. I gave away 8. How many do I have left?

I have 16 marbles. Sue has 7 fewer than I do. How many marbles does Sue have?

There are two serious flaws in the key word approach. First, these words may be used in many ways. They might be part of a problem that requires a different operation from the expected one:

There are 28 students in our class altogether. There are 13 boys. How many girls?

If we trust in key words, then *altogether* in this problem should signal addition of the numbers in the problem: 28 + 13, whereas the problem actually calls for finding the *difference* between 28 and 13.

The second reason for avoiding reliance on key words is that we want students to think through the entire structure of the problem. They need to read the problem and understand the situation so that they can construct a model of the problem for themselves. Here's another example:

I want to make cookies for my party. There will be 6 people at my party, including me. I want each person to have 4 cookies. How many cookies should I make altogether?

If students are encouraged to use key words, they are likely to pull numbers out of the problem and carry out some operation—in this case, perhaps 6 + 4—without developing a model of the whole problem, a structure of 6 equal groups of 4.

INVESTIGATION 4

One Hundred

What Happens

Session 1: Exploring the 100 Chart As a class, students fill in missing numbers on the 100 chart. In pairs they look for patterns on the 100 chart and write statements about the patterns they notice.

Sessions 2, 3, and 4: Working with 100 Students locate Today's Number on the 200 chart and calculate how far it is from 100 and other important numbers, such as 50 and the nearest multiple of 10. They are introduced to a new choice, which focuses on accumulating 100 cubes and organizing them into a 10-by-10 square. This, along with a variation of Close to 20 and a selection of addition and subtraction story problems, is a Choice Time activity.

Session 5: Penny-a-Pocket Students collect data about the number of pockets worn by class members. Each pocket is worth one penny. Students work individually to calculate how many pennies their pockets are worth. They record their strategy for solving this problem using words, numbers, and drawings. This activity is used as an assessment. When students are finished they continue working on choices from the previous days.

Mathematical Emphasis

- Becoming familiar with the structure of 100
- Working with 100 as a quantity
- Using the 100 chart as a tool for combining and comparing numbers
- Using familiar addition combinations to find totals
- Developing strategies to solve addition and subtraction problems

What to Plan Ahead of Time

Materials

- Crayons or markers (Session 1)
- Hundred Number Wall Chart with number cards 1–100 and transparent pattern markers (Sessions 1–4)
- Interlocking cubes: class set (Sessions 2–4)
- Number cubes with numbers or dots 1–6: 2 per 2–3 students (Sessions 2–4)
- Number cards: 1 deck per pair and extras for homework (Sessions 2–4)
- Envelopes such as legal size or large brown: 7 (Sessions 2–4)
- Paste or glue sticks (Sessions 2–4)
- Pennies: about 100–150 (Session 5)
- Large jar (Session 5)
- Class list of names: 1 per student (Session 5)
- Overhead projector (Session 5, optional)

Other Preparation

For Session 1

- Duplicate Student Sheet 22, Side by Side 100 Charts, 1 per student plus extras.

For Sessions 2, 3, and 4

- Gather decks of number cards with wild cards from previous investigations. Prepare additional decks, as needed, by duplicating Student Sheets 2–5, Number Cards.
- Duplicate Student Sheet 10, Close to 20 Score Sheet, and Student Sheet 22, Side by Side 100 Charts, 1 of each per student.

- Prepare decks of Roll-a-Square Cards by duplicating (on oaktag for durability) Student Sheets 23 and 24, Roll-a-Square Cards, 1 deck per group of 2–3 students.
- If you have a blank 10-by-10 grid available with squares that match the size of interlocking cubes, duplicate the grid to be used as a gameboard for Roll-a-Square.
- Duplicate Student Sheet 25, Story Problems Set E (or copies of the problems you have created), 1 per student.
- Prepare the story problems on Student Sheet 25 by cutting apart the student sheet into individual problems. Store each problem in a separate envelope marked with its number. Paste an example of each problem on the front of the envelope so students can see which problem they are choosing.

For Session 5

- Duplicate Student Sheet 26, Penny-a-Pocket, 1 per student.
- Prepare and duplicate a class list of names, 1 per student.

Exploring the 100 Chart

Materials

- Hundred Number Wall Chart with number cards 1–100 and transparent pattern markers
- Crayons or markers
- Student Sheet 22 (1 per student plus extras)

What Happens

As a class, students fill in missing numbers on the 100 chart. In pairs they look for patterns on the 100 chart and write statements about the patterns they notice. Their work focuses on:

- constructing a 100 chart
- becoming familiar with the structure and patterns on the 100 chart

Start-Up

Today's Number   Students express Today's Number using pennies, nickels, dimes, or quarters. For example, if Today's Number is 58, possible combinations are: 25¢ + 25¢ + 5¢ + 1¢ + 1¢ + 1¢ or 10¢ + 10¢ + 10¢ + 10¢ + 10¢ + 5¢ + 1¢ + 1¢ + 1¢. Add a card to the class counting strip and fill in another number on the blank 200 chart.

Activity

Filling in the 100 Chart

Before beginning this activity, remove all the cards except 1–10 and a few of the multiples of 10 (such as 30, 50, 60, 80) from the Hundred Number Wall Chart. Display it where all students can see it and reach it. Distribute the remaining cards to each student.

I removed most of the cards from our chart. Can you help me put the chart back together? Each of you has a few numbers that belong in the chart. Raise your hand if you think you know which pocket one of your numbers belongs in.

As students place their numbers in the chart, ask them how they knew where to place them. Some students may not volunteer to place their numbers until enough of the sequence has been filled in, and other students will begin to see patterns of columns and rows and place their numbers based on these and offer strategies such as, "I know the 40's are four rows down and then I just count across."

Note how individual students are placing their cards. Do they count from 1 or use another familiar number to insert their card? Do they use patterns as a way of locating the pocket for their number?

If students place a card in the incorrect place, let this stand. Other students may notice that it is in the wrong place and comment, or when more cards have been placed in the chart, students will probably be able to see that in fact it is out of place and make the necessary adjustment.

If it seems as though students are familiar with the structure of the chart, call a couple of students at a time to insert the rest of their numbers to complete the chart.

The Hundred Number Wall Chart is a model for organizing numbers and can become a very useful tool for students to use.

Activity

What Do You Notice About the 100 Chart?

When the Hundred Number Wall Chart is complete, ask students to look at it for a few minutes.

Look carefully at the Hundred Number Wall Chart for a minute. Now turn to someone sitting near you and tell him or her one thing that you notice about the chart.

Ask one or two students to share their observations with the class. These observations may range from descriptions such as, "The numbers go from 1 to 100." to particular patterns such as, "All the counting-by-10 numbers go down one side."

As students share, record their observations in words on the chalkboard. Using the transparent pattern marker (or some other marker), highlight each observation on the chart. Since students will be writing down their observations and highlighting what they see on individual 100 charts, ask for only one or two observations. Show students Student Sheet 22.

Working with a partner, tell each other the things you have noticed about the Hundred Number Wall Chart. Each of you is going to record your observations on one of these 100 charts. Write your observation underneath the chart and then, using a crayon or a marker, highlight what you noticed on the 100 chart. You and your partner might choose to write about the same pattern, but you explain what you see using your own words.

Distribute a copy of Student Sheet 22, Side by Side 100 Charts, to each student. Encourage partners to discuss their observations before they begin to write about them. Partners may choose to write about the same or different things. What's most important is that they record what they see and then highlight that on the chart.

❖ **Tip for the Linguistically Diverse Classroom** Students might record any patterns they notice by writing the number pattern itself: 10, 20, 30, 40, and so on. Students can then continue to show the pattern visually by using a crayon to highlight it on the chart.

As you circulate around the classroom, take special note of the types of observations students are making. In preparation for the discussion at the end of this session, select a few students' observations that you want to have presented to the class. Students probably will have observations in common. Consider choosing a pattern that was noticed by many students. Though their highlighted charts will look the same, how they describe what they observed will probably be different.

When students have completed both 100 charts, ask them to cut apart the two charts and continue on a new sheet.

Five to 10 minutes before the end of the session, call the group together so that they can share their observations of the 100 chart. After each student shares, ask the group if anyone else highlighted that same pattern. Have each student read his or her observations. In doing this, students will see that there are a variety of ways to describe the same observation.

After students have shared, have them place their charts in a group on the rug or chalkboard. You may want to post these 100 charts around the classroom, organizing them according to the patterns that students noticed.

At the end of this discussion, tell students that sometimes 100 charts can be used as a tool to help solve problems just as the calculator and the cubes are tools for problem solving. Keep a supply of 100 charts available for students to use.

Mention to students that tomorrow during Choice Time, filling in the Hundred Number Wall Chart will be a choice for a small group of students at a time.

Working with 100

What Happens

Students locate Today's Number on the 200 chart and calculate how far it is from 100 and other important numbers, such as 50 and the nearest multiple of 10. They are introduced to a new choice, which focuses on accumulating 100 cubes and organizing them into a 10-by-10 square. This, along with a variation of Close to 20 and a selection of addition and subtraction story problems, is a Choice Time activity. Their work focuses on:

- using the 100 chart as a tool for comparing numbers
- comparing two numbers and figuring the difference
- using familiar combinations such as doubles and 10's to add strings of numbers
- solving addition and subtraction problems

Start-Up

Today's Number   Students express Today's Number using pennies, nickels, dimes, or quarters. For example, if Today's Number is 58, possible combinations are: 25¢ + 25¢ + 5¢+ 1¢ + 1¢ + 1¢ or 10¢ + 10¢ + 10¢ + 10¢ + 10¢ + 5¢ + 1¢ + 1¢ + 1¢. Add a card to the class counting strip and fill in another number on the blank 200 chart.

Materials

- Interlocking cubes (class set)
- Student Sheet 22 (1 per student, as needed)
- Prepared decks of Roll-a-Square Cards (1 deck per group of 2–3 students)
- Number cubes (2 per 2–3 students)
- 10-by-10 grid with each square about 1" (1 per student, optional)
- Number cards with wild cards (1 deck per pair and extras for homework)
- Student Sheet 10 (2 per student, 1 for classwork, 1 for homework)
- Student Sheet 11 (1 per student, homework)
- Hundred Number Wall Chart with number cards 1–100
- Prepared envelopes containing Student Sheet 25 (1 problem per envelope)

Activity

Identify Today's Number on the 200 chart by highlighting it with a piece of colored transparent plastic or marker. If you are not keeping track of the number of school days, choose a number in the 60's to highlight and have your class work with during the first part of this session. (The number 64 is used as an example here.)

Today we have been in school for 64 days. How far is it from 64 to 70? How could you solve this problem?

Today's Number: How Far from 100?

Now I'd like you to find how far it is from 64 to 100. Work on this problem with a partner or by yourself. Use any materials you want. When everyone has finished, we will share our problem-solving strategies.

Have materials such as interlocking cubes and 100 charts (Student Sheet 22) available for students to use. As students are working, circulate and jot down notes about how students approach this problem. Notice the variety of strategies and tools they use. If students are having difficulty, suggest a smaller problem, such as how far Today's Number is from 80. Similarly, if some students finish early, choose another problem for them to work on such as how far Today's Number is from 50 or from 25.

After 10 minutes call students together to discuss the problem. The **Dialogue Box,** How Far from 64 to 100? (p. 136), illustrates some of the ways second graders have solved this problem.

Activity

Roll-a-Square

Explain to students that for the remainder of this session and the next two sessions they will have a choice of activities to work on. One of these choices is a new game called Roll-a-Square.

To play Roll-a-Square, you roll two number cubes, add the numbers, then snap together that many interlocking cubes to form a flat square. Each row of your square should have 10 cubes in it.

Once a row has 10 cubes you can start a new row underneath. After you roll and take your interlocking cubes, pick a card from the deck of Roll-a-Square Cards. Sometimes these cards will ask you a question and sometimes these cards will tell you to take more cubes or give some back.

Demonstrate how to roll the number cubes, collect the right number of cubes, and snap them together into rows of 10. After each turn, choose a card and read the instruction aloud to the class.

My card says, "How many more cubes do you need to finish another row of 10?" I have 12 cubes now, one row of 10 and 2 cubes in the next row. How many more cubes do I need to finish this second row of 10?

You might want to read a few more cards aloud so that students understand that choosing a card and answering the question are important parts of this game. Encourage students to listen to how their partners answers each question to make sure they agree with their answer. If they do not, suggest that they consult another pair of students for help before they ask you.

When you have collected 100 cubes, your game is over. After you finish each game, please unsnap your square and place the cubes back into the bucket so that other students can play.

Since you will need 100 cubes per student in order to play this game, students will have to take turns playing. For this reason this activity is a choice along with two other activities.

❖ **Tip for the Linguistically Diverse Classroom**   As students play the game, pair English-proficient students with second-language learners. The students proficient at English can read the cards aloud and model the tasks, if necessary, for their partners.

Choice Time

For the remainder of Sessions 2, 3, and 4, students work on the four choices listed below. Limit the number of students at the 100 Chart and at Roll-a-Square activities because materials will be limited. Encouraging students to play only one game of Roll-a-Square at a time may help to rotate students through the choices. You might also want to use various techniques to help students know when it is their turn at each Choice Time activity. For example, when one group of students has completed the goal in Roll-a-Square, they can be responsible for letting another group know that it is their turn.

> 1. Roll-a-Square
>
> 2. Close to 20 with Wild Cards
>
> 3. 100 Chart
>
> 4. Story Problems

Choice 1: Roll-a-Square

Materials: Interlocking cubes (100 per student); number cubes (2 per 2–3 students); decks of Roll-a-Square Cards (Student Sheets 23 and 24)

A group of two or three players take turns rolling the number cubes and collecting that number of interlocking cubes. They snap the cubes together into rows of 10. The object is to form a 10-by-10 square. After each roll a player draws a card from the deck of Roll-a-Square Cards and follows the direction on the card. The game is over when each player completes a 10-by-10 square.

Note: If you have 10-by-10 grids with squares that match the size of your cubes, you can use them as game boards.

Choice 2: Close to 20 with Wild Cards

Materials: Decks of number cards, with 4 wild cards; Student Sheet 10, Close to 20 Score Sheet; interlocking cubes

Students will be familiar with the game Close to 20, which they played in Investigation 2. Introduce them to a variation of this game, which includes using four wild cards in the deck. In this version, wild cards can be used as any number.

Choice 3: 100 Chart

Materials: Hundred Number Wall Chart and number cards 1–100

Limit the number of students at this activity to four or fewer. Students remove all the cards from the chart, mix them up, and then try to reassemble the chart. Suggest that they deal out the cards to each player and take turns placing a card into the chart.

Choice 4: Story Problems

Materials: Prepared envelopes containing story problems cut from Student Sheet 25 (or copies of problems you have created), paste, cubes

Addition and subtraction story problems are organized in individual envelopes. Students choose and solve one problem at a time. Four problems can be pasted on a sheet of paper (two per side). For each problem, they record their strategy using words, numbers, and/or pictures and stories. Make sure that students solve problems 2 and 5. These two problems will be the basis for a discussion at the end of Session 4.

Observing the Students

During these Choice Time activities you may want to spend time with small groups of students in order to work with those who are having difficulty or with students you have not observed or worked with for a while.

Use the following questions to guide your observations as students are working:

Roll-a-Square

- How do students add the total on the number cubes?
- Are they able to answer questions on the cards?
- How do they calculate how far away they are from a multiple of 10?
- Ask students how many cubes they have collected. Are they able to use groups of 10 to figure the total, or do they count all from 1?

- Can they figure out how many more cubes they need to complete their 10-by-10 square?

Close to 20 with Wild Cards

- Do students select their three numbers randomly, or use strategies such as choosing all the highest numbers or choosing two high numbers and then comparing that total to 20?
- How do students use the wild card? Do they try to make 20 with the number cards they have so that they can save the wild card for another round? Do they add two numbers, compare the total to 20, then decide on the value of the wild card?

100 Chart

- How do students place cards into the chart? Do they go in order, or are they able to use the structure of the chart to place a card?
- Do they recognize and use patterns on the chart as they place their numbers?

Story Problems

- What strategies are students using to solve addition problems? Are they using groups in a meaningful way, or do they count by 1's? Are they able to break apart numbers into more familiar components and then find the total?
- For subtraction problems, are students able to model the problem using materials?
- Do they know which number represents the whole quantity and which part they are taking away? Do they take away by 1's or by groups?
- How do students record their problem-solving strategies?
- Do they have an idea whether their answer is reasonable?

At the end of each session, after students have cleaned up their materials, remind them to record what they have done on their Weekly Logs.

Class Discussion: Many Ways to Solve a Problem

Students should have completed Problems 2 and 5 on Student Sheet 25 during Choice Time. You may want to check, and if necessary, have students complete those problems now.

Leave about 20 minutes at the end of Session 4 for a class discussion. To prepare for this discussion, look through students' work and choose three or four students who have solved these problems using different strategies or materials. One purpose of this discussion is to highlight that there are many ways to solve the same problem.

Gather the class together, making sure they have their work with them. Present problem 2 to the class: **Jake had 36 fish in his tank. He gave 11 fish to Kira. How many fish are left in Jake's tank?**

Was this a problem where you had to combine or add numbers together, or was it a problem where you were separating or subtracting one part of a number from another? How did you know what to do?

Briefly have students share how they interpreted the problem and then move on to sharing strategies.

As students share, record their solutions on the board. If students used materials, have them demonstrate for the class how they used them to solve the problem. After each solution, ask the group who else used a similar strategy. In this way students need to consider how their strategy was similar to or different from someone else's. Instead of having every student share, you can acknowledge the work of many by asking students to raise their hands if they solved the problem in a similar way and then move on to "Who solved the problem in a different way?"

Ask a few students to share the story problems they wrote for problem 5, 33 + 16. Record two or three of their problems on the board. Discuss what about the problems lets them know that they have to combine the two numbers. Students can then share strategies for solving 33 + 16.

Note: Throughout the school year, present students with combining and separating situations either in the form of written problems or by giving them notation and having them write problems for each other to solve. Also look for opportunities in the classroom where adding or subtracting numbers naturally occurs.

Much of the work presented in this unit is followed up and extended in the grade 2 *Investigations* unit *Putting Together and Taking Apart*. If you are following the suggested grade 2 sequence, this unit is taught during the second half of the school year. Students will, however, benefit from having consistent opportunities to work on the addition and subtraction strategies they have begun to develop in this unit.

Sessions 2, 3, and 4 Follow-Up

Students play Close to 20 with wild cards with someone at home. Each student will need Close to 20 Score Sheet, Close to 20 (Directions), and a deck of number cards. Students may already have a deck of cards at home, but have extra decks available for those who need them.

 **Homework**

How Far from 64 to 100?

Thinking flexibly about numbers is an important skill in developing number sense. Just as we use familiar landmarks to find our way around a map or territory, we can also use familiar numbers such as multiples of 10 as landmarks in the number system. Tools such as the 100 chart can help students to visualize groups of 10 and distances between two numbers. For students who are not yet using groups of 10, the 100 chart provides a concrete model of the number system.

During the activity on p. 129, students were asked to find how far it is from 64 to 100, then share their strategies.

Bjorn: I just counted up on the 100 chart from 64 to 100 and I got 36. [*Bjorn demonstrates his strategy using the class 100 chart. He counts confidently as he touches each number from 64 to 100.*]

Lila: I said 64 to 70 is 6. Then 70 to 80 is 10. 80 to 90 is 10 more, and 90 to 100 is another 10. Then I added 6 + 10 + 10 + 10 and I got 36.

So Bjorn counted on by 1's and Lila used 10's and then 1's to get to 100.

Graham: I sort of did it like Lila, but once I got to 70 I just knew that it was 30 more to 100.

Ebony: I went back to 4 + 6 because I knew that was 10, so 64 + 6 would get you to the next 10. Then I just added up by 10's.

So you used combinations of 10 to help you. Thinking about a more manageable number can often help you think about a bigger number. Any other ideas?

Rosie: I counted up from 64 by 10's. I said 74, 84, 94 that's 30, then 95, 96, 97, 98, 99, 100 [*she keeps track on her fingers*] that's 6, so it's 36.

Lionel: Well, I did it like Bjorn, but I came out with 37.

Can you show us what you did on the 100 chart?

Lionel: I said 1, 2, 3, 4, 5, 6 . . . [*Lionel begins counting on 64. He points to the 64 and says "1," thus including it in his total.*] . . . 33, 34, 35, 36, 37. [*He ends emphatically on 100.*]

Bjorn: Lionel started counting on 64, and I started on 65.

Olga: I don't think you should include 64 as part of your count because you are already at 64 and you want to know how much more it is to 100. So the first number you count should be 65. I think that's why you got 37 and Bjorn got 36.

So deciding which number to start counting on is an important decision. You also could think of the problem like, "You have 64 marbles. How many more do you need to have 100 marbles?" Yes, Franco?

Franco: I have one more way. I went backward to 60 because 60 to 100 is 40. Then I took away 4 because it's really 64, and that was 36.

Ebony: Hey, I get it. It's like what I did. Except that I went up to 70 and you went back to 60! I'll try that next time.

Penny-a-Pocket

What Happens

Students collect data about the number of pockets worn by class members. Each pocket is worth one penny. Students work individually to calculate how many pennies their pockets are worth. They record their strategy for solving this problem using words, numbers, and drawings. This activity is used as an assessment. When students are finished they continue working on choices from the previous days. Their work focuses on:

- collecting data
- adding strings of numbers
- communicating their problem-solving strategy in writing

Start-Up

Today's Number   Individually students generate as many ways as they can think of to express Today's Number. Collect these papers and compare them to other examples of Today's Number that students have completed during this unit. Add a card to the class counting strip and fill in another number on the blank 200 chart.

Materials

- Pennies (100–150)
- Class list of names (1 per student)
- Glass jar
- Student Sheet 26 (1 per student)
- Materials such as interlocking cubes, 100 charts (optional)
- Overhead projector (optional)

Activity

Assessment

Collecting Pocket Data

The following data-collection activity involving the number of pockets people are wearing is a progression from the activity in Investigation 1, Session 11. Here students solve the problem individually, rather than as a whole class. This activity can be used as an assessment.

Today we are going to collect information about the number of pockets people are wearing. Instead of using cubes to keep track of the number of pockets, we'll use pennies, a penny for every pocket you are wearing.

As students individually count their pockets, pass around a container of pennies and have them take one penny for each pocket.

Distribute a class list of names to each student. Explain that for this activity each student will need to record how many pockets people are wearing. Beginning at the top of the list (it is easier to keep track this way), call on one student at a time.

Carla, how many pockets are you wearing today? Please put your pennies in the jar. Everyone should write a 4 next to Carla's name. OK, Paul, you're next. How many pockets do you have?

On the overhead or chalkboard record each student's name and pocket data. This will serve as a master list of the data. Once all the data have been recorded and the pennies have been collected in the jar, explain to students that each of them should use the information that they have recorded on their class list to figure out the total number of pockets worn by students in the class.

For this activity I would like to get an idea about how each of you is thinking about this problem and about the strategies that you are using. Lots of times we solve problems with a partner, but today I would like each of you to work alone.

Distribute Student Sheet 26, Penny-a-Pocket, to each student.

When you find how many pennies our pockets are worth, write about how you solved the problem on your Penny-a-Pocket recording sheet. If you used materials, please explain which ones you used and write about how you used them to help you solve this problem.

For this activity, students should have access to a variety of materials, with the exception of calculators. Because you want to assess how students are adding strings of numbers—if they are using doubles, combinations of 10's, or related combinations—and how they add larger numbers, this information may be less clear if they use a calculator.

As students are working, circulate around the room. As part of your assessment, you might want to jot down notes on a class list about how each student is approaching this problem. These observations, along with a student's work, can provide a more complete picture of each student.

As you observe students working, keep the following questions in mind:

- Are students grouping numbers together, or are they adding on each number in succession?
- Do students use addition combinations they know?
- What materials are students using? Do they use them from the beginning or do they add up all the combinations of numbers that they know and then use materials to deal with combining two-digit numbers?
- How do students combine two-digit numbers? Do they count on? add 10's? use materials?
- How do students organize and keep track of their work?

As you observe you may find that for some students you have to modify this problem. Suggest that these students choose a group of five friends (or some other appropriate number) and find the total number of pockets for this smaller group.

When students have finished, remind them that part of this task is also to communicate to others how they solved the problem. You may need to prompt and encourage students to write an explanation that is complete and that communicates their problem-solving process. For example, a student might write: "I added up the numbers and there were 88 pennies." Your response could be, "That's a good start, but your explanation doesn't tell me how you added up the numbers. Did you use materials? Which numbers did you add first and why?" Suggest that students who are having difficulty "talk through" their process with you as a way of clarifying their thinking.

Extending the Problem   For students who finish early you can extend the problem in the following ways:

- What's the fewest number of coins you could trade the pennies for?
- How many more pennies would you need to make $1?
- Compare the number of pockets we have this week to the number of pockets we had last week.

Looking at Student Work   As you evaluate students' solutions, keep the following questions in mind:

- How did students figure out the total number of pockets?
- Can students record their work clearly to communicate how they solved the problem?

At some point during this math class or later in the day, count the pennies in the jar and have a brief discussion about how students solved this problem.

Depending on how much opportunity your students have had playing Roll-a-Square and Close to 20 with wild cards, you may want to offer these as choices for the remainder of this session. You may also decide to spend one or two extra sessions on these activities so that students gain more experience with these games and other games in this unit.

Choosing Student Work to Save

As the unit ends, you may want to use one of the following options for creating a record of students' work.

- Students look back through their folders and think about what they learned in this unit, what they remember most, or what was hard or easy for them. You might have students discuss this with partners or share in the whole group.

- Depending on how you organize and collect student work, you may want to have students select some examples of their work to keep in a math portfolio. In addition, you may want to choose some examples from each student's folder to include. Items such as Today's Number, the Magic Pot assessment, the Penny-a-Pocket assessment, a combining problem, a separating problem, and 100 chart observations can be useful pieces for assessing student growth over the school year. You may want to keep the original and make copies of these pieces for students to take home.

- Send a selection of work home for families to see. Students write a cover letter describing their work in this unit. This work should be returned if you are keeping a year-long portfolio of math work for each student.

Today's Number

Today's Number is one of three routines that are built into the grade 2 *Investigations* curriculum. Routines provide students with regular practice in important mathematical ideas such as number combinations, counting and estimating data, and concepts of time. For Today's Number, which is done daily (or most days), students write equations that equal the number of days they have been in school. Each day, the class generates ways to make that number. For example, on the tenth day of school, students look for ways to combine numbers and operations to make 10.

This routine gives students an opportunity to explore some important ideas in number. By generating ways to make the number of the day, they explore:

- number composition and part-whole relationships (e.g., 10 can be $4 + 6$, $5 + 5$, or $20 - 10$)
- equivalent arithmetical expressions
- different operations
- ways of deriving new numerical expressions by systematically modifying prior ones (e.g., $5 + 5 = 10$, so $11 = 5 + 6$)

Students' strategies evolve over time, becoming more sophisticated as the year progresses. Early in the year, second graders use familiar numbers and combinations, such as $5 + 5 = 10$. As they become accustomed to the routine, they begin to see patterns in the combinations and have favorite kinds of number sentences. Later in the year, they draw on their experiences and increased understanding of number. For example, on the forty-ninth day they might include $100 - 51$, or even $1000 - 951$ in their list of ways to make 49. The types of number sentences that students contribute over time can provide you with a window into their thinking and their levels of understanding of numbers.

If you are doing the full-year grade 2 curriculum, Today's Number is introduced in the first unit, *Mathematical Thinking at Grade 2*. Throughout the curriculum, variations are often introduced as whole-class activities and then carried on in the Start-Up section. The Start-Up section at the beginning of each session offers suggestions of variations and extensions of Today's Number.

While it is important to do Today's Number every day, it is not necessary to do it during math time. In fact, many teachers have successfully included Today's Number as part of their regular routines at the beginning or end of each day. Other teachers incorporate Today's Number into the odd 10 or 15 minutes that exist before lunch or before a transition time.

If you are teaching an *Investigations* unit for the first time, rather than using the number of days you have been in school as Today's Number, you might choose to use the calendar date. (If today is the 16th of the month, 16 is Today's Number.) Or you might choose to begin a counting line that does not correspond to the school day number. Each day, add a number to the strip and use this as Today's Number. Begin with the basic activity and then add variations once students become familiar with this routine.

The basic activity is described below, followed by suggested variations.

Materials

- Chart paper
- Student Sheet 1, Weekly Log
- Interlocking cubes

If you are doing the basic activity, you will also need the following materials:

- Index cards (cut in half and numbered with the days of school so far, e.g., 1 through 5 for the first week of school)
- Strips of adding-machine tape
- Blank 200 chart (tape two blank 100 charts together to form a 10-by-20 grid)

Continued on next page

Basic Activity

Initially, you will want to use Today's Number in a whole group, starting during the first week of school. After a short time, students will be familiar with the routine and be ready to use it independently.

Establishing the Routine

Step 1. Post the chart paper. Call students' attention to the small box on their Weekly Logs in which they have been recording the number of the days they have been in school.

Step 2. Record Today's Number. Write the number of the day at the top of the chart paper. Ask students to suggest ways of making that total.

Step 3. List the number sentences students suggest. Record their suggestions on chart paper. As you do so, invite the group to confirm each suggestion or discuss any incorrect responses, and to explain their thinking. You might have cubes available for students to double-check number sentences.

Step 4. Introduce the class counting strip. Show students the number cards you made and explain that the class is going to create a counting strip. Each day, the number of the day will be added to the row of cards. Post the cards in order in a visible area.

Step 5. Introduce the 200 chart. Display the chart and explain that another way the class will keep track of the days in school is by filling in the chart. Record the appropriate numbers in the chart. Tell the class that each day the number of the day will be added to the chart. To help bring attention to landmark numbers on the chart, ask questions such as, "How many more days until the tenth day of school? the twentieth day?"

Variations

When students are familiar with the structure of Today's Number, you can connect it to the number work they are doing in particular units.

Make Today's Number Ask students to use some of the following to represent the number:

- only addition
- only subtraction
- both addition and subtraction
- three numbers
- combinations of 10 ($23 = 4 + 6 + 4 + 6 + 3$ or $23 = 1 + 9 + 2 + 8 + 3$)
- a double ($36 = 18 + 18$ or $36 = 4 + 4 + 5 + 5 + 9 + 9$)
- multiples of 5 and 10 ($52 = 10 + 10 + 10 + 10 + 10 + 2$ or $52 = 5 + 15 + 20 + 10 + 2$)

Use the idea of working backward. Put the number sentences for Today's Number on the board and ask students to determine what day you are expressing: $10 + 3 + 5 + 7 + 5 + 4 = ?$ Notice how students add this string of numbers. Do they use combinations of 10 or doubles to help them?

In addition to defining how Today's Number is expressed, you can vary how and when the activity is done:

Start the Day with Today's Number Post the day's chart paper ahead of time. When students begin arriving, they can generate number sentences and check them with partners, then record their ways to make the number of the day before school begins. Students can review the list of ways to make the number at that time or at the beginning math class. At whole-group meeting or morning meetings, add the day's number to the 200 chart and the counting strip.

Choice Time Post chart paper with the Number of the Day written on it so that it is accessible to students. As one of their choices, students generate number sentences and check them with partners, then record them on the chart paper.

Continued on next page

Work with a Partner   Each student works with a partner for 5 to 10 minutes and lists some ways to make the day's number. Partners check each other's work. Pairs bring their lists to the class meeting or sharing time. Students have their lists of number sentences in their math folders. These can be used as a record of students' growth in working with number over the school year.

Homework   Assign Today's Number as homework. Students share number sentences sometime during class the following day.

Catch-Up   It can be easy to get a few days behind in this routine, so here are two ways to catch up. Post two or three Number-of-the-Day pages for students to visit during Choice Time or free time. Or assign a Number of the Day to individual students. Each can generate number sentences for his or her number as well as collect number sentences from classmates.

Class History   Post "special messages" below the day's number card to create a timeline about your class. Special messages can include birthdays, teeth lost, field trips, memorable events, as well as math riddles.

Today's Number Book   Collect the Today's Number charts in a *Number of the Day Book*. Arrange the pages in order, creating chapters based on 10's. Chapter 1, for example, is ways to make the numbers 1 through 10, and combinations for numbers 11–20 becomes Chapter 2.

How Many Pockets?

How Many Pockets? is one of three classroom routines presented in the grade 2 *Investigations* curriculum. Routines provide students with regular practice in important mathematical ideas such as number combinations, counting and estimating data, and concepts of time. In How Many Pockets? students collect, represent, and interpret numerical data about the number of pockets everyone in the class is wearing on a particular day. This routine often becomes known as Pocket Day. In addition to providing opportunities for comparison of data, Pocket Days provide a meaningful context in which students work purposefully with counting and grouping. Pocket Day experiences contribute to the development of students' number sense—the ability to use numbers flexibly and to see relationships among numbers.

If you are doing the full-year grade 2 curriculum, we suggest that you collect pocket data at regular intervals throughout the year. Many teachers collect pocket data every tenth day of school.

The basic activity is described below, followed by suggested variations. Variations are introduced within the context of the *Investigations* units. If you are not doing the full grade 2 curriculum, we suggest that you begin with the basic activity and then add variations when students become familiar with this routine.

Materials

- Interlocking cubes
- Large jar
- Large rubber band or tape
- Hundred Number Wall Chart and number Cards (1–100)
- Pocket Data Chart (teacher-made)
- Class list of names
- Chart paper

1	2	3	4	5	6	7	8	9	10
11	12	13	14	15	16	17	18	19	20
21	22	23	24	25	26	27	28	29	30
31	32	33	34	35	36	37	38	39	40
41	42	43	4	45	46	47	48	49	50
51	52	53	54	55	56	57	58	59	60
61	62	63	64	65	66	67	68	69	70
71	72	73	74	75	76	77	78	79	80
81	82	83	84	85	86	87	88	89	90
91	92	93	94	95	96	97	98	99	100

Hundred Number Wall Chart

How many pockets are we wearing today?		
	Pockets	People
Pocket Day 1		

Pocket Data Chart

Basic Activity

Step 1. Students estimate how many pockets the class is wearing today. Students share their estimates and their reasoning. Record the estimates on chart paper. As the Pocket Days continue through the year, students' estimates may be based on the data recorded on past Pocket Days.

Step 2. Students count their pockets. Each student takes 1 interlocking cube for each pocket he or she is wearing.

Continued on next page

Step 3. Students put the cubes representing their pockets in a large jar. Vary the way you do this. For example, rather than passing the jar around the group, call on students with specific numbers of pockets to put their cubes in the jar (e.g., students with 3 pockets). Use numeric criteria to determine who puts their cubes in the jar (e.g., students with more than 5 but fewer than 8 pockets).

Step 4. With students, agree on a way to count the cubes. Count the cubes to find the total number of pockets. Ask students for ideas about how to double-check the count. By re-counting, students see that a group of objects can be counted in more than one way; for example, by 1's, 2's, 5's, and 10's. With many experiences, they begin to realize that some ways of counting are more efficient than others and that a group of items can be counted in ways other than by 1, without changing the total.

Primary students are usually most secure counting by 1's, and that is often their method of choice. Experiences with counting and grouping in other ways help them begin to see that number is conserved or remains the same regardless of its arrangement—20 cubes is 20 whether counted by 1's, 2's, or 5's. Students also become more flexible in their ability to use grouping, especially important in our number system, in which grouping by 10 is key.

Step 5. Record the total for the day on a Pocket Data Chart. Maintaining a chart of the pocket data as they are accumulated provides natural opportunities for students to see that data can change over time and to compare quantities.

How many pockets are we wearing today?		
	Pockets	People
Pocket Day 1	41	29

Variations

Comparing Data Students revisit the data from the previous Pocket Day and the corresponding cube level marked on the now empty jar.

On the last Pocket Day, we counted [*give number*] pockets. Do you think we will be wearing more, fewer, or about the same number of pockets today? Why?

After students explain their reasoning, continue with the basic activity. When the cubes have been collected, invite students to compare the level of cubes now with the previous level and to revise their estimates based on this visual information.

Discuss the revised estimates and then complete the activity. After you add the day's total to the Pocket Data Chart, ask students to compare and interpret the data. To facilitate discussion, build a train of interlocking cubes for today's and the previous Pocket Day's number. As students compare the trains, elicit what the cube trains represent and why they have different numbers of cubes.

Use the Hundred Number Wall Chart Do the basic activity, but this time let students choose only one way to count the cubes. Then introduce the Hundred Number Wall Chart as a tool that can be used for counting cubes. This is easiest when done with students sitting on the floor in a circle.

To check our pocket count, we'll put our cubes in the pockets on the chart. A pocket can have just one cube, so we'll put one cube in number 1's pocket, the next cube in number 2's pocket, and keep going in the same way. How many cubes can we put in the first row?

Students will probably see that 10 cubes will fill the first row of the chart.

One group of 10 cubes fits in this row. What if we complete the second row? How many rows of the chart do you think we will fill with the cubes we counted today?

Continued on next page

Encourage students to share their thinking. Then have them count with you and help to place the cubes one by one in the pockets on the chart. When finished, examine the chart together, pointing out the total number of cubes in it and the number of complete rows. For each row, snap together the cubes to make a train of 10. As you do so, use the rows to encourage students to consider combining groups of 10. Record the day's total on your Pocket Data Chart.

Note: If cubes do not fit in the pockets of the chart, place the chart on the floor and place the cubes on top of the numbers.

Find the Most Common Number of Pockets
Each student connects the cubes representing his or her pockets into a train. Before finding the total number of pockets, sort the cube trains with students to find the most common number of pockets. Pose and investigate additional questions too, such as:

- **How many people are wearing the greatest number of pockets?**

- **Is there a number of pockets no one is wearing?**

- **Who has the fewest pockets?**

The cubes are then counted to determine the total number of pockets.

Take a Closer Look at Pocket Data Each student builds a cube train representing his or her pockets. Beginning with those who have zero pockets, call on students to bring their cube trains to the front of the room. Record the information in a chart, such as the one shown here. You might make a permanent chart, with blanks for placing number cards.

Pose questions about the data, such as, "Two people each have 2 pockets. How many pockets is that?" Then record the number of pockets.

_0_ people have 0 pockets.	_0_ pockets
_4_ people have 1 pocket.	_4_ pockets
_2_ people have 2 pockets.	_4_ pockets
_2_ people have 3 pockets.	_6_ pockets

To work with combining groups, you might keep a running total of pockets as data are recorded in the chart until you have found the cumulative total.

We counted 12 (for example) **pockets, and then we counted 6 pockets. How many pockets have we counted so far? Be ready to tell us how you thought about it.**

As students give their solutions, encourage them to share their mental strategies. Alternatively, after all data have been collected, students can work on the problem of finding the total number of pockets.

Graphing Pocket Data Complete the activity using the variation Finding the Most Common Number of Pockets. Leave students' cube trains intact. Each student then creates a representation of the day's pocket data. Provide a variety of materials, including stick-on notes, stickers or paper squares, markers and crayons, drawing paper, and graph paper for students to use.

These cube trains represent how many pockets people are wearing today. Suppose you want to show our pocket data to your family, friends, or students in another classroom. How could you show our pocket data on paper so that someone else could see what we found out about our pockets today?

Continued on next page

By creating their own representations, students become more familiar with the data and may begin to develop theories as they consider how to communicate what they know about the data to an audience. Students' representations may not be precise; what's important is that the representations enable them to describe and interpret their data.

Compare Pocket Data with Another Class
Arrange ahead of time to compare pocket data with a fourth or fifth grade class. Present the following question to students:

Do you think fifth grade students wear more, fewer, or about the same number of pockets as second grade students? Why?

Discuss students' thinking. Then investigate this question by comparing your data with data from another classroom. One way to do this is to invite the other class to participate in your Pocket Day. Do the activity first with the second graders, recording how many people have each number of pockets on the Pocket Data Chart and finding the total number of pockets. Repeat with the other students, recording their data on chart paper. Then compare the two sets of data.

How does number of pockets in the fifth grade compare to the number of pockets in second grade? Why?

Discuss students' ideas.

Calculate the Total Number of Pockets   Divide students into groups of four or five. Each group determines the total number of pockets being worn by the group. Data from each small group are shared and recorded on the board. Using this information, students work in pairs to determine the total number of pockets worn by the class. As a group, they share strategies used for determining the total number of pockets.

In another variation, students share individual pocket data with the group. Each student records this information using a class list of names to keep track. They then determine the total number of pockets worn by the students in the class. Observe how students calculate the total number of pockets. What materials do they use? Do they group familiar numbers together such as combinations of 10, doubles, or multiples of 5?

Time and Time Again

Time and Time Again is one of three classroom routines included in the grade 2 *Investigations* curriculum. This routine helps students develop an understanding of time-related ideas such as sequencing of events, the passage of time, duration of time periods, and identifying important times in their day.

Because many of the ideas and suggestions presented in this routine will be incorporated throughout the school day and into other parts of the curriculum, we encourage teachers to use this routine in whatever way meets the needs of their students and their classroom. We believe that learning about time and understanding ideas about time happen best when activities are presented *over* time and have relevance to students' experiences and lives.

Daily Schedule   Post a daily schedule. Identify important times (start of school, math, music, recess, reading) using both analog (clockface) and digital (10:15) representation. Discuss the daily schedule each day and encourage students to compare the actual starting time of, say, math class with what is posted on the schedule.

Talk Time   Identify times as you talk with students. For example, "In 15 minutes we will be cleaning up and going to recess." Include specific times and refer to a clock in your classroom, "It's now 10:15. In 15 minutes we will be cleaning up and going out to recess. That will be at 10:30."

Timing One Hour   Set a timer to go off at one-hour intervals. Choose a starting time and write both the analog time (use a clockface) and the digital time. When the timer rings, record the time using analog and digital times. At the end of the day, students make observations about the data collected. Initially you'll want to use whole and half hours as your starting points. Gradually you can use times that are 10 or 20 minutes after the hour and also appoint students to be in charge of the timer and of recording the times.

Timing Other Intervals   Set a timer to go off at 15-minute intervals over a period of two hours. Begin at the hour and after the data have been collected discuss with students what happened each time 15 minutes was added to the time (11:00, 11:15, 11:30, 11:45). You can also try this with 10-minute intervals.

Home Schedule   Students make a schedule of important times at home. They can do this both for school days and for nonschool days. They should include both analog and digital times on their schedules. Later in the year they can use this schedule to see if they were really on time for things like dinner, piano lessons, or bedtime. They record the actual time that events happened and calculate how early or late they were. Students can illustrate their schedules.

Comparing Schedules   Partners compare important times in their day, such as what time they eat dinner, go to bed, get up, leave for school. They can compare whether events are earlier or later, and some students might want to calculate how much earlier or later these events occur.

Life Line   Students create a timeline of their life. They interview family members and collect information about important developmental milestones such as learning to walk, first word, first day of school, first lost tooth, and important family events. Students then record these events on a life line that is a representation of the first seven or eight years of their life.

Clock Data   Students collect data about the types of clocks they have in their home—digital or analog. They make a representation of these data and as a class compare their results.

- **Are there more digital or analog clocks in your house?**
- **Is this true of our class set of data?**
- **How could we compare our individual data to a class set of data?**

Continued on next page

Time Collection   Students bring in things from home that have to do with time. Include digital and analog clocks as well as timers of various sorts. These items could be sorted and grouped in different ways. Some students may be interested in investigating different types of time-pieces such as sundials, sand timers and pendulums.

How Long Is a Minute (or half a minute)?   As you time 1 minute, students close their eyes and then raise their hands when they think a minute has gone by. Ask, "Is a minute longer or shorter than you imagined?" Repeat this activity or have students do this with partners.

What Can You Do in a Minute?   When students are familiar with timing 1 minute, they work in pairs and collect data about things they can do in 1 minute. Brainstorm a list of events that students might try. Some ideas that second graders have suggested include: writing their name, doing jumping jacks or sit-ups, hopping on one foot, saying the ABC's, snapping together interlocking cubes, writing certain numbers or letters (this is great practice for working on reversals), and drawing geometric shapes such as triangles, squares, or stars. Each student chooses four or five activities to do in 1 minute. Before they collect the data, they predict how many they can do in 1 minute. Then with partners they gather the data and compare.

How Long Does It Take?   Using a stopwatch or a clock with a second hand, time how long it takes students to complete certain tasks such as lining up, giving out supplies, or cleaning up after math time. Emphasize doing these things in a responsible way. Students can take turns being "timekeepers."

Stopwatches   Most second graders are fascinated by stopwatches. You will find that students come up with many ideas about what to time. If possible, acquire a stopwatch for your classroom. (Inexpensive ones are available through educational supply catalogs.) Having stopwatches available in the classroom allows students to teach each other about time and how to keep track of time.

The following activities will help ensure that this unit is comprehensible to students who are acquiring English as a second language. The suggested approach is based on *The Natural Approach: Language Acquisition in the Classroom* by Stephen D. Krashen and Tracy D. Terrell (Alemany Press, 1983.) The intent is for second-language learners to acquire new vocabulary in an active, meaningful context.

Note that *acquiring* a word is different from *learning* a word. Depending on their level of proficiency, students may be able to comprehend a word upon hearing it during an investigation, without being able to say it. Other students may be able to use the word orally but not read or write it. The goal is to help students naturally acquire targeted vocabulary at their present level of proficiency.

We suggest using these activities just before the related investigations. The activities can be led by English-proficient students.

Investigation 2

imagine, story problem

1. Draw a picture of a tree on the chalkboard or on chart paper. Ask students to close their eyes and picture the tree, then *imagine* a bird in the tree. With their eyes remaining shut, ask students to describe the colors of their bird. Once students open their eyes, ask them if they see a bird on the tree drawn on the chalkboard. Explain that the bird they saw was in their *imagination,* so only they could see it.

2. Draw a swing hanging from the tree. Ask students to imagine a child in the swing and describe something about the child (for example, boy or girl, age, what child is wearing).

3. Ask students to keep their eyes closed as you tell this story.

   Imagine that there is a boy on the swing. He is swinging high into the air. Now imagine that two more children are walking toward the boy. They are coming over to watch him swing. How many children can you see now?

Ask students to open their eyes and describe the number of children. Write on the chalkboard: 1 + 2 = 3. Then ask them to tell how this number sentence relates to the children in the story problem.

pennies, nickels, dimes, quarters, coins, group

1. Give each student a penny, nickel, dime, and quarter. Use action statements to help familiarize students with the coins.

   Put the penny in your hand.

   Put the nickel next to the dime.

   Hold your quarter up over your head.

   Put all your coins in a pile.

2. Use action statements in which students identify coins by their value.

   Can you show me a coin that is worth 5 cents?

   Can you show me a coin that is worth 10 cents?

   Can you show me which coin is worth the most money?

Blackline Masters

Family Letter
Student Sheet 1, Weekly Log
Student Sheet 2, Number Cards (0–2)
Student Sheet 3, Number Cards (3–5)
Student Sheet 4, Number Cards (6–8)
Student Sheet 5, Number Cards (9–10)
Student Sheet 6, Tens Go Fish (Directions)
Student Sheet 7, Turn Over 10 (Directions)
Student Sheet 8, Number Strings
Student Sheet 9, More Number Strings
Student Sheet 10, Close to 20 Score Sheet
Student Sheet 11, Close to 20 (Directions)
Student Sheet 12, Beat the Calculator Cards
Student Sheet 13, Beat the Calculator Cards
Student Sheet 14, Our Class and the Magic Pot
Student Sheet 15, People and Pet Riddles
Student Sheet 16, Counting on Our Fingers
Student Sheet 17, Shop and Save
Student Sheet 18, Story Problems, Set A
Student Sheet 19, Story Problems, Set B
Student Sheet 20, Story Problems, Set C
Student Sheet 21, Story Problems, Set D
Student Sheet 22, Side by Side 100 Charts
Student Sheet 23, Roll-a-Square Cards
Student Sheet 24, More Roll-a-Square Cards
Student Sheet 25, Story Problems, Set E
Student Sheet 26, Penny-a-Pocket

_____ , 19 _____

Dear Family,

In math class, we are beginning a unit on the number system called *Coins, Coupons, and Combinations.* In this unit, your child will learn about how numbers are made from other numbers—20 can be made from 10 and 10 or from four 5's or from ten 2's. Being able to take numbers apart and put them back together flexibly is the basis for developing good number sense.

First we will work with addition combinations, exploring combinations of 10 (4 + 6, 2 + 8) and doubles (4 + 4, 5 + 5). We will then use these addition combinations to learn others. The goal is for children to become familiar with number combinations through repeated use and by learning about relationships among numbers.

In the second half of the unit your child will be working with numbers such as 5, 10, 15, 20, 25, 50, and 100. Activities include using coins to find different ways to make 25¢ and figuring out how to save 50¢ at the grocery store using combinations of coupons. Students will also work on story problems for which they will use addition and subtraction to solve.

While our class is working on this unit you can help in several ways:

■ For homework, your child will be bringing home materials and directions for games we play at school. Many of these involve work with number combinations. Please help your child find a safe place to store these materials and directions since some of them will be used repeatedly throughout the unit. An empty file folder or manila envelope might make a convenient storage place.

■ If you have a penny jar at home suggest that your child count out a handful of pennies. In school they are encouraged to check their count by counting a second time in a different way (by 2's or 5's). Students might compare their handful of pennies to your handful or to a brother's or sister's handful.

■ Ask your child to count the change in your pocket. We have been working mostly with pennies, nickels, and dimes, so you might want to have them count only these coins at first.

■ If your child goes to the grocery store with you and you are using a coupon, point out how much each coupon is worth and if possible show your child the amount using coins. Eventually your child can be in charge of the coupons for your family!

■ We will be using coupons from the newspaper or from the mail in some of our activities. Please send in any store coupons that you won't be using.

Thanks for your help.

Sincerely,

Weekly Log

Day Box

Monday, _____

Tuesday, _____

Wednesday, _____

Thursday, _____

Friday, _____

Number Cards (0-2)

0	0	0	0
1	1	1	1
2	2	2	2

Number Cards (3–5)

3	3	3	3
4	4	4	4
5	5	5	5

Number Cards (6–8)

6	6	6	6
7	7	7	7
8	8	8	8

Number Cards (9–10)

9	**9**	**9**	**9**
10	**10**	**10**	**10**
Wild Card	Wild Card	Wild Card	Wild Card

Tens Go Fish

Materials: Deck of number cards 0–10 (four of each) with wild cards removed

Players: 3 to 4

How to Play

The object of this game is to get two cards that total 10.

1. Each player is dealt five cards. The rest of the cards are placed facedown in the center of the table.

2. If you have any pairs of cards that total 10, put them down in front of you and replace those cards with cards from the deck.

3. Take turns. On a turn, ask <u>one</u> other player for a card that will go with a card in your hand to make 10.

4. If you get a card that makes 10, put the pair of cards down. Take one card from the deck. Your turn is over.
   If you do not get a card that makes 10, take the top card from the deck. Your turn is over.

   If the card you take from the deck makes 10 with a card in your hand, put the pair down and take another card.

5. If there are no cards left in your hand but still cards in the deck, you take two cards.

6. The game is over when there are no more cards.

7. At the end of the game, make a list of the number pairs you made.

Turn Over 10

Materials: Deck of number cards 0–10 (four of each) plus four wild cards

Players: 2 to 3

How to Play

The object of the game is to turn over and collect combinations of cards that total 10.

1. Arrange the cards facedown in four rows of five cards. Place the rest of the deck facedown in a pile.

2. Take turns. On a turn, turn over one card and then another. A wild card can be made into any number.

    If the total is less than 10, turn over another card.

    If the total is more than 10, your turn is over and the cards are turned facedown in the same place.

    If the total is 10, take the cards and replace them with cards from the deck. You get another turn.

3. Place each of your card combinations of 10 in separate piles so they don't get mixed up.

4. The game is over when no more 10's can be made.

5. At the end of the game, make a list of the number combinations for 10 that you made.

Number Strings

$5 + 7 + 5$	$1 + 6 + 9$
$8 + 9 + 1$	$7 + 3 + 9$
$4 + 3 + 2 + 1$	$9 + 7 + 1 + 3$
$7 + 5 + 7 + 5$	$8 + 6 + 3 + 7 + 2$

More Number Strings

8 + 9 + 3	2 + 7 + 6
2 + 9 + 1 + 2	6 + 6 + 7 + 7
9 + 4 + 5 + 1	5 + 3 + 6 + 8 + 7
8 + 6 + 8 + 4	5 + 6 + 4 + 10 + 5

Close to 20 Score Sheet

GAME 1   SCORE

Round 1: _____ + _____ + _____ = _____   _____

Round 2: _____ + _____ + _____ = _____   _____

Round 3: _____ + _____ + _____ = _____   _____

Round 4: _____ + _____ + _____ = _____   _____

Round 5: _____ + _____ + _____ = _____   _____

                                TOTAL SCORE   _____

GAME 2   SCORE

Round 1: _____ + _____ + _____ = _____   _____

Round 2: _____ + _____ + _____ = _____   _____

Round 3: _____ + _____ + _____ = _____   _____

Round 4: _____ + _____ + _____ = _____   _____

Round 5: _____ + _____ + _____ = _____   _____

                                TOTAL SCORE   _____

Close to 20

Materials: Deck of number cards 0–10 (four of each) with the wild cards removed; Student Sheet 10, Close to 20 Score Sheet; counters

Players: 2 to 3

How to Play

The object of the game is to choose three cards that total as close to 20 as possible.

1. Deal five cards to each player.

2. Take turns. Use any three of your cards to make a total that is as close to 20 as possible.

3. Write these numbers and the total on the Close to 20 Score Sheet.

4. Find your score. The score for the round is the difference between the total and 20. For example, if you choose $8 + 7 + 3$, your total is 18 and your score for the round is 2.

5. After you record your score, take that many counters.

6. Put the cards you used in a discard pile and deal three new cards to each player. If you run out of cards before the end of the game, shuffle the discard pile and use those cards again.

7. After five rounds, total your score and count your counters. These two numbers should be the same. The player with the lowest score and the fewest counters is the winner.

Beat the Calculator Cards

5 + 4 + 1	3 + 3 + 3 + 1
9 + 2 + 10	1 + 1 + 1 + 1
4 + 5 + 6 + 3	4 + 10 + 4 + 2
6 + 6 + 12	10 + 9 + 1 + 10

Beat the Calculator Cards

$5 + 5 + 1 + 1$	$8 + 5 + 2$
$2 + 2 + 2 + 4$	$9 + 1 + 5$
$10 + 6 + 6$	$6 + 5 + 1$
$1 + 2 + 3 + 4$	$9 + 2 + 9$

Our Class and the Magic Pot

1. Suppose our class fell into the magic pot and doubled the number of people.

   How many people would there be?

   Write about how you could solve the problem. Use words, numbers, and pictures to explain your thinking.

2. What would be good about having a double class?

3. What would be hard about having a double class?

People and Pet Riddles

1. Read this riddle.

> There are 14 legs in this group.
> There are 6 heads in this group.
> There are 12 ears in this group.
> There are 50 fingers in this group.
> There is 1 tail in this group.

Who could be in this group?

2. Write your answer and draw a picture if you want to.

3. On another paper, write a riddle about the people and pets that live in your home.

Write your riddle on the front of your paper and the answer on the back. Tomorrow we will solve each other's riddles.

Counting on Our Fingers

How many fingers are there in our classroom?
Write about how you solved this problem.
Use numbers, words, or pictures to show your thinking.

There are _____ fingers in our classroom.

Shop and Save

1. Choose two coupons that total 50¢ exactly. List them here.

2. **It's Party Time**

   Choose three coupons for things you need for your party. List them. How much did you save?

3. **Time to Clean**

   How much money can you save on three things to clean with?

4. **Double Coupons!**

   Your store is doubling the value of coupons. The total of your coupons cannot be more than $1. Find two coupons that when doubled save you as close to $1 as possible.

5. Find three coupons that total 90¢ exactly. List them here.

6. **Lunch Box Food**

   Find three coupons for things you would like to have in your lunch box. List the coupons. How much can you save?

Story Problems
Set A

1. There are 12 children playing tag on the playground. Then 10 more children join the game. How many children are playing tag now?

---

2. There are 31 cars and 14 buses in the school parking lot. How many cars and buses are there in all?

Story Problems
Set B

1. Luis and Kathy were collecting rocks. Luis found 16 rocks and Kathy found 24. How many rocks did the children collect?

---

2. Tom collects stamps. For his birthday he got 29 stamps from his grandfather and 22 stamps from his mom. How many new stamps did Tom add to his collection?

---

3. Make up a story problem to go with 21 + 14. Then solve the story problem. Write and explain how you solved the problem.

Story Problems
Set C

1. Andy had 28 balloons.
By mistake he let go and 15 of them flew away.
How many balloons did Andy have left?

---

2. There are 48 pencils in a box. Every student in the
class got a new pencil. There are 27 students
in the class. How many pencils were left?

Story Problems
Set D

1. Kira and Jake were eating peanuts. There were 39 peanuts in the bag. Kira ate 18 of them and Jake ate the rest. How many peanuts did Jake eat?

2. Kira and Jake were making snowballs. They both made 19 snowballs. How many did they make in all?

3. There were 47 students in the gym. One class of 23 students went back to their room. How many students were left in the gym?

4. The second grade class went on a trip to the zoo. There were 32 students and 12 adults on the trip. How many people went on the zoo trip?

5. Marcel and Shawn were counting pennies. Marcel had 33 pennies and Shawn had 36 pennies. How many pennies did they have in all?

6. Lin has 50 pennies in a cup. She spent 29 pennies at the store on a pencil. How much money does Lin have left?

7. Make up a problem to go with 30 – 13.
   Write about how you solved the problem.
   Use words, numbers, or pictures to show your thinking.

Name

Side by Side 100 Charts

1	2	3	4	5	6	7	8	9	10
11	12	13	14	15	16	17	18	19	20
21	22	23	24	25	26	27	28	29	30
31	32	33	34	35	36	37	38	39	40
41	42	43	44	45	46	47	48	49	50
51	52	53	54	55	56	57	58	59	60
61	62	63	64	65	66	67	68	69	70
71	72	73	74	75	76	77	78	79	80
81	82	83	84	85	86	87	88	89	90
91	92	93	94	95	96	97	98	99	100

1	2	3	4	5	6	7	8	9	10
11	12	13	14	15	16	17	18	19	20
21	22	23	24	25	26	27	28	29	30
31	32	33	34	35	36	37	38	39	40
41	42	43	44	45	46	47	48	49	50
51	52	53	54	55	56	57	58	59	60
61	62	63	64	65	66	67	68	69	70
71	72	73	74	75	76	77	78	79	80
81	82	83	84	85	86	87	88	89	90
91	92	93	94	95	96	97	98	99	100

Roll-a-Square Cards

How many cubes do you have in all?	How many more cubes do you need to finish another row of 10?
How far from 50 cubes are you?	How many cubes do you have? TAKE 10 cubes. Now how many cubes do you have?
How many cubes do you need to add or take away so that you have 50 cubes in all?	Roll the number cubes again. TAKE double the number of cubes.

More Roll-a-Square Cards

How many cubes do you have in all?	How many more cubes do you need to finish another row of 10?
How many cubes do you have? GIVE BACK 10 cubes. Now how many cubes do you have?	TAKE 5 extra cubes. How many cubes do you have?
Roll the number cubes again. TAKE double the number of cubes.	TAKE another 10 cubes. How many cubes do you have?

Story Problems
Set E

1. Kira has a tank of fish. She has 15 goldfish and 24 guppies. How many fish are in Kira's tank?

2. Jake had 36 fish in his tank. He gave 11 fish to Kira. How many fish are left in Jake's tank?

3. The second grade is collecting cans for recycling. So far they have filled two bags. Each bag has 26 cans in it. How many cans have been collected?

4. Jamal and his dad made 42 cookies for Jamal's class. There are 20 children in Jamal's class. If each child eats 1, how many cookies will be left?

5. Write a story problem that goes with 33 + 16. Then solve the problem.

6. Write a story problem that goes with 43 − 31. Then solve the problem.

Penny-a-Pocket

1. Use the pocket data from our class to solve the Penny-a-Pocket problem.

The pockets in our class are worth _____ pennies.

---

2. Write about how you solved the problem. Use words, numbers, or pictures to show your thinking.

Include the class list and any other paper you used to solve this problem.